Design for Deflection

A Monograph covering:

- Introduction to Problem
- Mechanical Strain Energy
- Slender Member Strain Energy
- Castigliano's Theorem
- Example Applications
- Design Related Problems

Carl F. Zorowski
Design Engineering Monograph VI

Design for Deflection

Copyright 2017
All Rights Reserved

Design for Deflection

Table of Contents

Chapter 1 – Introduction

	Page
Introduction	1
Loaded Crank Handle	1
Classic Mechanics Approach	2
Handle Loadings and Deflections	2
Extension Loadings and Deflections	3
A Different and Easier Way	4
Example Mechanics Approach	4
A Curved Beam	6

Chapter 2 – Mechanical Strain Energy

Introduction	9
Force due to Normal Stress	9
Elongation due to Normal Stress	9
Work Done by Force dF	10
Tension Strain Energy	11
Force due to Shear Stress	11
Displacement due to Shear Stress	12
Work Done by Shear Force	13
Shear Strain Energy	13
Complete General Stress State	14
Elastic Material Properties	15
Generalized Hooke's Law	16
Strain Energy in Terms of Stresses	17

Special Case-Hydrostatic Pressure 18
Strain Energy- Plain Stress State 19

Chapter 3 – Slender Member Strain Energy and Castigliano's Theorem

Introduction 21
Common Loading Forms 21
Normal Stress Components 22
Shear Stress Components 23
Member Strain Energy 23
Simplified Total Energy 24
Castigliano's Theorem 26
Theorem Application 27
Cantilever Beam Example 28
Curved Beam Example 30

Chapter 4 – Application of Castigliano's Theorem

Introduction 33
Cantilever Beam Example 33
Uniformly Loaded Cantilever 35
Curved Beam Example 36
Curved Quarter Loop 41
Coil Spring Deflection 45
Statically indeterminate Beam 46
Beam Built in at Both Ends 48
Two Element Structure 52

Design for Deflection

Truss Deflection Example 55
Truss Deflection at F 57

Chapter 5 – Design Related Problems
Introduction 59
Prob. 1 – Cantilever Torsion Spring 59
Problem Definition 60
Total Strain Energy 61
Force – Moment Relationship 62
Effective Spring Constant 63
Geometric Approximation Error 64
Disk Equilibrium Correction 65
Magnitude of Error 66
Calculating Beam Dimensions 68
Prob. 2 – Lifting a Long Pipe 69
Problem Statement 70
Problem Definition 70
Free Body Diagram 71
Applying Equilibrium 71
Geometric Requirements 72
Implementing Requirements 73
F in terms of δ and L 74
M_o in terms of δ and L 74
Solution for Lifted Length L 75
Determining F, R and M_o 76
Numerical Values of L 77
Calculation of F and M_o 78
Shear and Bending Moment 78

Design for Deflection

Total Lifting Force F_t	80
Variation of F_t with δ	80
Design in Practical Units	81
Practical Design Values	82

Preface

Work done by an external force in deforming an elastic material is stored in the material as potential energy that can be recovered. This work is referred to as mechanical strain energy and can be calculated in terms of the stresses induced and the strains experienced by the material. It can also be related directly to the external loading that produces bending, torsion and extension of slender members. In this format it can be used to determine the deformation of the member by application of a unique mathematical theorem.

This monograph describes how this strain energy is calculated, it's format in loadings applied to slender members, how deformations are obtained with the mathematical theorem with examples and applications to two practical design problems.

Chapter 1 is an introduction to the problem of determining deformations in slender members by comparing the classical mechanics approach to the use of strain energy methods.

Chapter 2 develops how the work done by simple normal and shear stresses along with their respective strains contributes to the strain energy stored in a deformed body. Hooke's Law relationships for an elastic material are introduced to eliminate shear strains from the strain energy. It is further simplified to apply to a state of plane stress in preparation for inclusion of the classic slender member loading modes.

In Chapter 3 the normal and shear stresses in the plane state of stress expression for strain energy are replaced by the bending moment, torques and tensions in slender members that create the stresses. In this format Castigliano's theorem is introduced and applied to determine generically the deflections and slopes that exist at applied moments, torques and axial tension or compression.

Chapter 4 is devoted to examples that illustrate the application of strain energy and Castigliano's theorem to straight and curved members. Included are finding deflections and slopes where no forces or moments act, creating additional equations for statically indeterminate problems and analyzing multi member structures including planer trusses.

In Chapter 5 strain energy methods are used in detailed design related operational analyses of two practical problems. One deals with determining design dimensions of a slender member to satisfy specified design specifications. The second problem investigates deformation performance limitation in an elastic structure that behaves nonlinearly.

As in previous monographs in this "Design for xxx" series about one third of the total content is devoted to figures that consist of mathematical developments related to the subject content. It is

Design for Deflection

recommended that the reader follow and work through these developments along with the text. This will provide a better understanding of the principles involved, how they are used to achieve the desired final generic mathematical formulations and their detailed application to example problems.

The contents of this monograph made up a portion of a Mechanical Design Engineering course taught by the author in the NCSU "Engineering on Line" distance education graduate program.

Carl F. Zorowski
January 2017
Cary, NC

Design for Deflection

Design for Deflection

Chapter 1 – Introduction

This first chapter of Design for Deflection deals with comparing the classical mechanics approach to the use of energy methods in determining the deflections of slender members and structures subjected to bending, torsional and extension loading.

Loaded Crank Handle

Shown in the Figure 1-1 is a crank handle arrangement of two cantilevered beams subjected to a force F acting in the xy plane. The problem is to determine the deflection in the direction of the force F due to the bending of the handle section of the crank and the additive contributions of the torsion, bending and axial displacements in the section that connects the handle to its fixed position at the stationary wall.

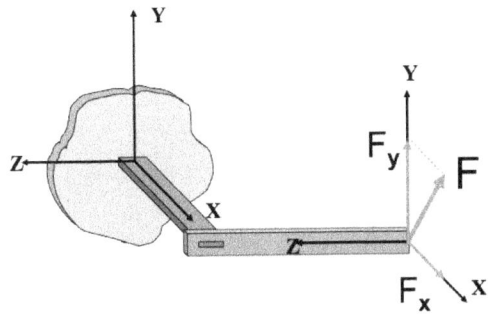

Figure 1-1 Loaded Crank Handle

1

Design for Deflection

Classic Mechanics Approach

The classical mechanics approach is to first separate the structure into easily analyzed units. The second step is to apply equilibrium to each of the separate units to determine the loading to which they are subjected. The deflections and/or deformations of each unit are then calculated using standard models that are beam bending, axial extension and torsional twist. Finally the deformations of all the separate units are combined appropriately to determine the total structure deflection.

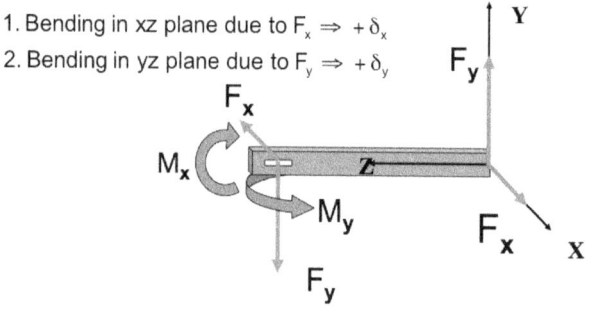

Figure 1-2 Handle Portion Loading/Deflection

Handle Loadings and Deflections

The components of the applied Force F, F_x and F_y, will generate bending deflections in the x and y directions at the right end of the beam pictured in Figure 1-2. These can be easily determined relative to the left end of the beam as δ_x and δ_y assuming the left end is fixed. Applying equilibrium to this section results in the requirement of reaction moments M_x and M_y together with forces F_x and F_y to exist at the

Design for Deflection

left end of the beam. These become the forces and moments the will produce deformations in the section of the structure that is fixed at the wall and will have to be added appropriately to the bending deflections δ_x and δ_y.

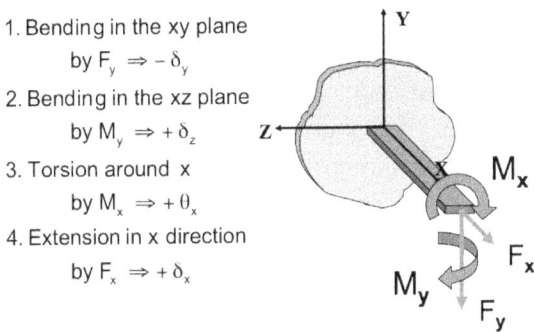

1. Bending in the xy plane
 by $F_y \Rightarrow -\delta_y$
2. Bending in the xz plane
 by $M_y \Rightarrow +\delta_z$
3. Torsion around x
 by $M_x \Rightarrow +\theta_x$
4. Extension in x direction
 by $F_x \Rightarrow +\delta_x$

Figure 1-3 Wall Extension Loading and Deflection

Wall Extension Loadings and Deflections

The forces and moments transmitted from the left end of the handle section of the structure to the cantilever fixed at the wall will result in four distinct deformations and deflections. The tensile force F_x will produce an axial extension δ_x. The moment M_y will result in a bending deflection in the xz plane of δ_z. The moment M_x will produce a torsional rotation θ_x about the x-axis and the force F_y will produce a bending deflection $-\delta_y$ in the y direction. Each of these will have to be determined independently and then appropriately added to the bending deflections already determined in the handle section.

Design for Deflection

A Different and Easier Way

As described in the previous several Figures using a classical mechanics approach to determine the deflection of a complex structure can be cumbersome and involved, providing many opportunities to make algebraic or other mathematical mistakes in the process. There is an easier and much more convenient way to solve this type of problem by the use of strain energy principles. This approach is based on the work done by the loading forces when the structure deflects and permits the inclusion of the effects of bending, torsion and extension along the entire structure in one equation. It is also particularly useful for determining the behavior of indeterminate structures.

Example Mechanics Approach

The efficiency of the strain energy approach compared to the classical mechanics solution can be demonstrated with the simple example of a cantilever beam shown in Figure 1-4 subjected to an end load F. To find the end deflection using a mechanics approach requires the solution of the differential equation $EI\, d^2y/dx^2 = -M(x)$, the bending moment. E is the modulus of elasticity of the material and I is the cross sectional moment of inertia of the beam. In this problem $M(x) = -Fx$ measured from the right end of the beam. Integrating the differential equation results in

$$EI\, y = Fx^3/6 + C_1 x + C_2.$$

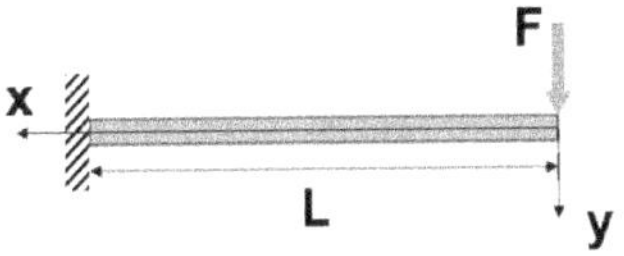

Design for Deflection

Find deflection due to F

apply $EI\dfrac{d^2y}{dx^2} = -M(x)$

with $M(x) = -Fx$

$(EI)\dfrac{d^2y}{dx^2} = Fx$

$(EI)\dfrac{dy}{dx} = F\dfrac{x^2}{2} + C_1$

$(EI)\, y = F\dfrac{x^3}{6} + C_1 x + C_2$

B.C. at $x = L$ $y = 0$, at $x = L$ $dy/dx = 0$

Figure 1-4 Example Mechanics Approach

The constants are determined from the conditions that at the built in end of the beam the deflection y and the slope dy/dx must both be zero. Applying the two boundary conditions in Figure 1-4 gives C1 = - FL²/2 and C2 = FL³/3. Evaluating the final equation with the calculated values of C_1 and C_2 inserted for y at x =0 results in the deflection of the right end of the beam as FL³/3EI in Figure 1-5.

Apply B.C

$(EI)\dfrac{dy}{dx} = 0 = F\dfrac{L^2}{2} + C_1$

$(EI)\, y = 0 = F\dfrac{L^3}{6} + C_1 L + C_2$

therefore

$C_1 = -\dfrac{FL^2}{2}$ and $C_2 = \dfrac{FL^3}{3}$

so that

$y = \dfrac{FL^3}{EI}\left(\dfrac{1}{6}\left(\dfrac{x}{L}\right)^3 - \dfrac{1}{2}\left(\dfrac{x}{L}\right) + \dfrac{1}{3}\right)$ at $x = 0$ $y = \dfrac{FL^3}{3EI}$

Figure 1-5 Solution Completed

Design for Deflection

Example – Strain Energy Method

Without explaining the details of the strain energy method of solution, to be covered in subsequent chapters, the problem of the end deflection of the cantilever beam is solved in three lines. The first equation in Figure 1-6 represents the total strain energy in terms of the bending moment as a function of Fx^2 integrated over the length of the beam. The second line gives the deflection at the point of application of the force F by differentiating the strain energy U partially by F. This is simply F/EI times the integral of $x^2\, dx$ over the length of the beam resulting in $y = FL^3/3EI$ the same result as obtained by the strength of materials approach. The simplicity of this approach is obvious even in this simple example.

Apply strain energy technique

$$U = \frac{1}{2EI}\int [M(x)]^2 dx = \frac{1}{2EI}\int_0^L (Fx)^2 dx$$

$$y = \frac{\partial U}{\partial F} = \frac{1}{2EI}\int_0^L 2(Fx)(x)dx = \frac{F}{EI}\int_0^L x^2 dx$$

$$y = \frac{FL^3}{3EI} \quad \text{(same result)}$$

Figure 1-6 Strain Energy Solution

A Curved Beam

Another convenient feature of the strain energy method is the ease with which it can also be used to analyze the deflections of curved beams which is virtually impossible by the classical strength of materials approach. In subsequent chapters of this monograph the theory of the strain energy method is developed and applied to a variety of example problems.

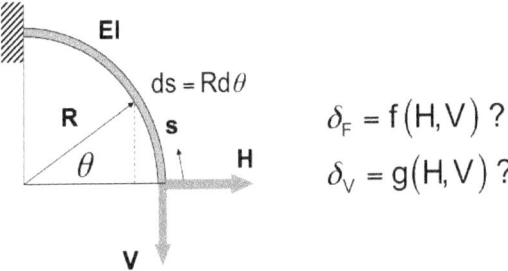

Figure 1-7 Curved Beam Problem

Design for Deflection

Design for Deflection

Chapter 2 – Mechanical Strain Energy

Introduction

Prior to applying the strain energy method to determine deflections it is first necessary to develop a general expression for mechanical strain energy produced by normal and shear stresses in an elastic material.

Force due to Normal Stress

Shown in Figure 2-1 is an incremental cubic element of dimensions dx, dy, dz acted on by a normal tensile stress σ_x on the dz dy face of the element in the x direction. This gives rise to an incremental force dF in the x direction of σ_x dy dz.

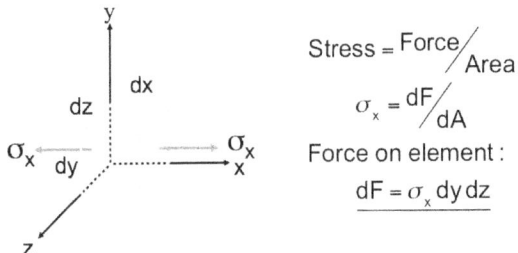

Figure 2-1 *Force due to Normal Stress*

Elongation due to Normal Stress

A normal strain ε_x is defined as the ratio of an elongation δ in the direction of x divided by the original length L over which the elongation took place. The normal stress σ_x produces an incremental elongation of the cubical element in the x direction.

Design for Deflection

This incremental elongation dδ is expressed as ε_x dx as illustrated in Figure 2-2.

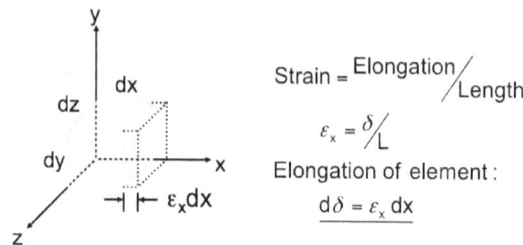

Figure 2-2 Elongation due to Normal Stress

Work Done by Force dF

Elongation increases linearly with the application of force in an elastic material. The work done by the force acting through this elongation is the area under the curve of force versus elongation as shown in Figure 2-3.

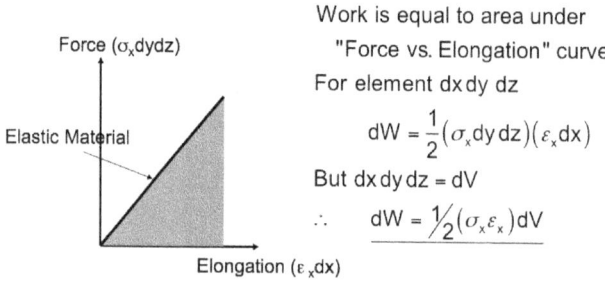

Figure 2-3 Work Done by Force dF

For the cubical element under consideration the incremental work dW is expressed as one half the product of the incremental force, $\sigma_x\,dydz$, multiplied by the incremental elongation, $\varepsilon x\,dx$. This can be rewritten as one half the product of $\sigma_x\varepsilon_x\,dV$ where dV is the incremental volume $dxdydz$.

Tension Strain Energy

When work is done by a force in deforming an elastic material, like the stretching of a spring in tension, the energy equivalent of the work performed is stored in the material and can be recovered. This stored energy is referred to as strain energy of deformation and is represented by dU. For an elastic material the incremental strain energy dU is equal to the incremental work dW. The applied tension on the cubical element and its subsequent extension results in an increment of strain energy dU equal to $\frac{1}{2}(\sigma_x\varepsilon_x)dV$, see Figure 2-4.

$$dW(\text{work}) = dU(\text{strain energy})$$

For tension in the x direction on cubical element

$$dU = \frac{1}{2}(\sigma_x\varepsilon_x)dV$$

Figure 2-4 Tension Strain Energy

Force due to Shear Stress

In Figure 2-5 a shear stress τ_{xy} applied to the top face of the cubical element will result in an incremental shear force that can be expressed as dF_s equal to $\tau_{xy}\,dxdz$. This force will produce a shearing displacement of the top surface in the x direction.

Design for Deflection

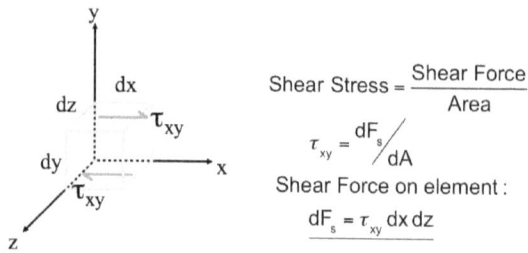

Figure 2-5 Force due to Shear Stress

Shear Displacement due to Shear Force

Shear strain is defined classically as angular change per unit dimensional length. For the cubical element it is the angle between the original orientation of the dy side of the element and its distorted orientation due to the shearing effect of the applied shear stress on area dx dz. It is designated γx_y as illustrated in Figure 2-6.

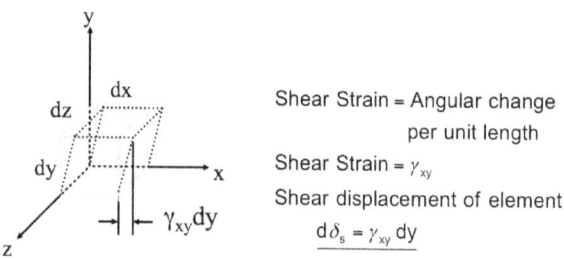

Figure 2-6 Shear Displacements from Shear Stress

The shear displacement of the top surface of the element through which the shear force will act can be expressed as $d\delta_s$ equal to γ_{xy} dy.

Design for Deflection

Work done by Shear Force

For an elastic material the shear displacement will increase linearly with an increase in the shear force as depicted in Figure 2-7. The work done by this shear force is the area under the curve. For the cubical element it is 1/2 τ_{xy} dx dz, multiplied by the shear displacement γ_{xy} dy. This can be rewritten as ½ τ_{xy} γ_{xy} dV.

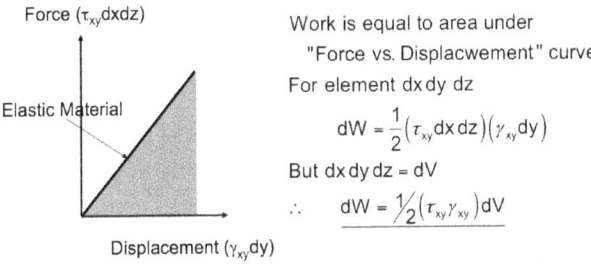

Figure 2-7 Work done by Shear Force

Shear Strain Energy

The work done by a shearing force that produces shear deformation in an elastic body also results in strain energy being stored in the material that can be recovered similarly to strain energy produced by normal stress and strain.

The incremental strain energy dU that is equal to the incremental work dW resulting from the shear stress τx_y that produces the shear stain γx_y is expressed as 1/2 τ_{xy} γ_x dV as shown in Figure 2-8.

Design for Deflection

dW (work) = dU(strain energy)
For shear in the x direction on cubical element
$$dU = \tfrac{1}{2}(\tau_{xy}\gamma_{xy})dV$$

Figure 2-8 Shear Strain Energy

Complete General Stress State

A body subjected to a complete general state of stress can be represented by a cubicle element on which three normal stresses, σ_x, σ_y and σ_z act on its three mutually perpendicular faces. In addition there will also exist three shear stresses, τ_{xy}, τ_{xz} and τ_{yz} acting on these same three mutually perpendicular faces. These six components of stress define the complete state of stress at any point in the body as shown in Figure 2-9.

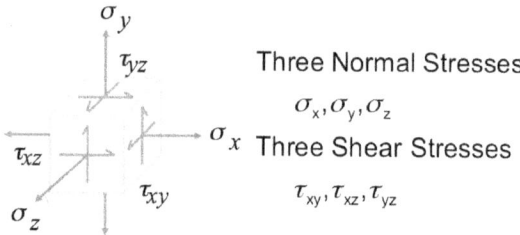

Three Normal Stresses
$\sigma_x, \sigma_y, \sigma_z$
Three Shear Stresses
$\tau_{xy}, \tau_{xz}, \tau_{yz}$

Figure 2-9 Complete General Stress State

Total Strain Energy per Unit Volume

Each one of the six components of a general stress state together with their respect strains produce a contribution to the total strain energy of the body. This is expressed as the total incremental strain energy dU for a unit volume of dV in Figure 2-10 as one half the sum of the product of the normal stresses and normal strains, $\sigma_x\varepsilon_x$ plus $\sigma_y\varepsilon_y$ and $\sigma_z\varepsilon_z$, times dV plus one half the sum of the shear stress and shear strains, $\tau_{xy}\gamma_{xy}$ plus $\tau_{xz}\gamma_{xz}$ and $\tau_{yz}\gamma_{yz}$, times dV. Whether all of these normal and shear stress exist in any specific instance is dependent on the external loading of the structural element.

$$dU = \frac{1}{2}\left(\sigma_x\varepsilon_x + \sigma_y\varepsilon_y + \sigma_z\varepsilon_z\right)dV + \frac{1}{2}\left(\tau_{xy}\gamma_{xy} + \tau_{xz}\gamma_{xz} + \tau_{yz}\gamma_{yz}\right)dV$$

Figure 2-10 Total Strain Energy per Unit Volume

Elastic Material Properties

By introducing the three material property constants for a homogeneous isotropic elastic media the general equation for incremental strain energy can be reformulated. This results in a more convenient format for considering the effects of the bending, torsion and extension loadings generally applied to slender members. In Figure 2-11 the modulus of elasticity or Young's modulus defined as the slope of the normal stress strain curve or stress divided by strain is designated E. In a similar fashion the shear modulus of the material is designated as G. The third material constant is Poisson's ratio represented by the Greek letter μ. It physically represents the lateral contraction that takes place perpendicular to the direction of extension produced by an applied tension.

Design for Deflection

Definitions:

$$E = \sigma_i / \varepsilon_i \quad E = \text{Young's Modulus}$$
$$(\text{Modulus of Elasticity})$$
$$G = \tau_{ij} / \gamma_{ij} \quad G = \text{Shear Modulus}$$
$$\varepsilon_j = -\mu \varepsilon_i = -\mu / E \sigma_i$$
$$\mu = \text{Poisson's Ratio}$$

Figure 2-11 Elastic Material Properties

Generalized Hooke's Law

As a consequence of the Poisson ratio effect normal strains are a function of all three normal stresses. This results in ε_x in Figure 2-12 as 1/E times the quantity $[\sigma_x - \mu(\sigma_y + \sigma_z)]$. Even if σ_x is zero there would still be strain in the x direction if either σ_y or σ_z or both existed.

Complete general Stress – Strain relations for an elastic material including Poison's ratio effect

$$\varepsilon_x = \frac{1}{E}\left(\sigma_x - \mu(\sigma_y + \sigma_z)\right) \quad \gamma_{xy} = \frac{\tau_{xy}}{G}$$
$$\varepsilon_y = \frac{1}{E}\left(\sigma_y - \mu(\sigma_x + \sigma_z)\right) \quad \gamma_{yz} = \frac{\tau_{yz}}{G}$$
$$\varepsilon_z = \frac{1}{E}\left(\sigma_z - \mu(\sigma_y + \sigma_x)\right) \quad \gamma_{xz} = \frac{\tau_{xz}}{G}$$

Figure 2-12 Generalized Hooke's Law

No interaction of the shear strains takes place so t γ_{xy} is simply τ_{xy}/G. The formulation of the normal strain equations for ε_y and ε_z follow directly from that of ε_x. The same is true for the other two remaining shear strain components

Strain Energy in Terms of Stresses

The normal and shear strain components can now be eliminated from the general equation for strain energy by introducing the Hooke's law relationships from Figure 2-12. By collecting similar types of stress forms the final expression for the total strain energy is expressed in Figure 2-13. It consists of three distinct stress groupings that include normal stress components squared, products of normal stress and shear stress components squared.

Eliminate strains from total strain energy by substituting Hooke's law relations

$$dU = \left\{ \frac{1}{2E} \left[\sigma_x \left(\sigma_x - \mu (\sigma_y + \sigma_z) \right) + \ldots \right] \right\} dV + \left\{ \frac{1}{2G} \left[\tau_{xy} (\tau_{xy}) + \ldots \right] \right\} dV$$

or finally

$$dU = \left\{ \frac{1}{2E} \left(\sigma_x^2 + \sigma_y^2 + \sigma_z^2 \right) - \frac{\mu}{E} \left(\sigma_x \sigma_y + \sigma_y \sigma_z + \sigma_x \sigma_z \right) \right\} dV + \left\{ \frac{1}{2G} \left(\tau_{xy}^2 + \tau_{xz}^2 + \tau_{yz}^2 \right) \right\} dV$$

Figure 2-13 Strain Energy in Terms of Stresses

Special Case – Hydrostatic Pressure

An interesting application of the general strain energy equation that relates to the properties of elastic media is to subject a unit cubic element to hydrostatic pressure. In this instance the three normal stress σ_x, σ_y and σ_z are set equal to -p, the hydrostatic pressure. All three shear stress components are set equal to zero.

Design for Deflection

Substituting the stress state that defines hydrostatic pressure into the general equation for strain energy in terms of stress components leads to the incremental stain energy dU equal to (1/2E)(3p²) minus (µ /E) (3p²) all multiplied by dV as shown in Figure 2-14. Integrating this over the unit cube and collecting like terms gives U, the total stain energy, equal to (3p²/2E)(1-2µ). Since the total strain energy can never be negative this means that µ must always be between 0 and 1/2 for an elastic material. For essentially all metals this is true in the linear region of their stress strain behavior.

$$dU = \left\{ \frac{1}{2E}\left(\sigma_x^2 + \sigma_y^2 + \sigma_z^2\right) - \frac{\mu}{E}\left(\sigma_x\sigma_y + \sigma_y\sigma_z + \sigma_x\sigma_z\right)\right\}dV$$

$$+ \left\{\frac{1}{2G}\left(\tau_{xy}^2 + \tau_{xz}^2 + \tau_{yz}^2\right)\right\}dV$$

Set $\sigma_x = \sigma_y = \sigma_z = -p$ and $\tau_{xy} = \tau_{xz} = \tau_{yz} = 0$
then

$$dU = \left\{\frac{1}{2E}(3p^2) - \frac{\mu}{E}(3p^2)\right\}dV$$

$$U = \frac{3p^2}{2E}(1 - 2\mu) \text{ and since } U \geq 0$$

then for an elastic material $0 < \mu < \frac{1}{2}$

Figure 2-14 Special Case – Hydrostatic Pressure

Strain Energy –Plain Strain State

The state of plain strain is of particular interest in the bending, torsion and extension of slender members. This special state of stress is described by the restrictions:

$$\sigma_z = \mu(\sigma_x + \sigma_y) \text{ and } \tau_{xz} = \tau_{yz} = 0.$$

Design for Deflection

These constraints are substituted into the general equation for an increment of strain energy dU in Figure 2-15. Collecting like terms and simplifying results in dU = $(1/2E)(\sigma_x^2 + \sigma_y^2 - \mu\,\sigma_x\sigma_y) + (1/2G)(\tau_{xy}^2)$. In Chapter 3 this strain energy equation will be further developed to permit it application in terms of bending moments, twisting torques and axial forces applied to loaded slender members.

$$dU = \left\{\frac{1}{2E}\left(\sigma_x^2 + \sigma_y^2 + \sigma_z^2\right) - \frac{\mu}{E}\left(\sigma_x\sigma_y + \sigma_y\sigma_z + \sigma_x\sigma_z\right)\right\}dV$$
$$+ \left\{\frac{1}{2G}\left(\tau_{xy}^2 + \tau_{xz}^2 + \tau_{yz}^2\right)\right\}dV$$

Set $\sigma_z = \mu(\sigma_x + \sigma_y)$ and $\tau_{xz} = \tau_{yz} = 0$ then

$$dU = \left\{\frac{1}{2E}\left(\sigma_x^2 + \sigma_y^2 + \mu^2(\sigma_x + \sigma_y)^2\right) - \frac{\mu}{E}\left(\sigma_x\sigma_y + \mu(\sigma_x + \sigma_y)^2\right)\right\}dV$$
$$+ \left\{\frac{1}{2G}\left(\tau_{xy}^2\right)\right\}dV$$

Expanding, collecting and simplfying gives

$$dU = \left\{\frac{1}{2E}\left(\sigma_x^2 + \sigma_y^2 - \mu\sigma_x\sigma_y\right) + \frac{1}{2G}\tau_{xy}^2\right\}dV$$

Figure 2-15   Strain Energy – Plane Strain

Design for Deflection

Chapter 3 – Slender Member Strain Energy and Castigliano's Theorem

Introduction

Chapter 3 introduces the classical slender member loadings of bending, torsion and axial forces into the strain energy equation for a state of plane strain. Also covered are Castigliano's theorem and its use with strain energy to determine slender member deflection behavior.

Common Loading Forms

Depicted in Figure 3-1 is a slender member with the three major loading types, bending, torsion and axial extension, experienced at an internal cross section as a consequence of some external loading. For convenience in the development that follows, bending is represented by two moment components, M_y and M_z, about their respective axes.

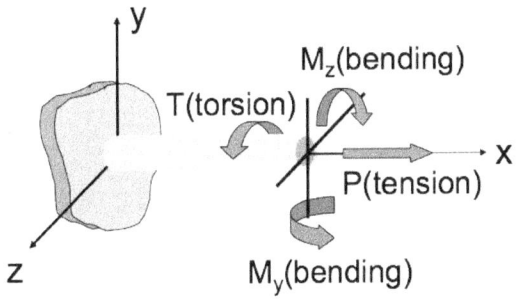

Figure 3-1 Common Loading Forms

A shear force V almost always exists internally but it effects on lateral deflections will be neglected

Design for Deflection

since it contribution only becomes important in members whose length is short compared to transverse dimensions.

Normal Stress Components

Bending moments and axial loading give rise to normal stress components σ_x on cross sections perpendicular to the axis of the member. The normal stress component σ_x due to the axial load assuming a uniform distribution is simply the force P in the axial direction divided by the cross sectional area A as shown in Figure 3-2. There are two normal stress components σ_x generated by the bending moments about the y and z axes. The σ_x distribution due to M_z is given by (M_z/ I_z) y. In a similar fashion the σ_x distribution due to M_y is (M_y/ I_y) z. Recall that these two expressions are only valid if I_y and I_z are *principal moments of inertia*. In dealing with structural members that are long compared to their lateral dimensions the normal stress σ_y can be taken to be zero.

Tension Load in x direction

$$\sigma_x^T = P/A \quad A = \text{cross sectional area}$$

Bending about z and y axes

$$\sigma_x^{B_z} = \left(M_z/I_z\right)y, \quad \sigma_x^{B_y} = \left(M_y/I_y\right)z$$

where I_z and I_y must be principal area moments of inertia

and

$$\sigma_y = 0$$

Figure 3-2 Normal Stress Components

Design for Deflection

Shear Stress Components

For torsion of circular cross section shafts the shear stress distribution is given by the shear stress $\tau = Tr/J$, where r is the radius measured from the center of the shaft and J is the polar moment of inertia of the cross section. The radius r is related to y and z coordinates in the yz plane by the expression $r^2 = y^2 + z^2$. It is recognized that shear stress τ is in reality the shear component τ_{yz} so that τ_z^2 becomes $T^2(y^2 + z^2)$ squared in Figure 3-3. For a circular cross section shaft under the loading assumed τ_{xy} and τ_{xz} will both be zero.

Torsion about x axis (circular cross section)

$$\tau = \frac{Tr}{J} \quad \text{but} \quad r^2 = y^2 + z^2 \quad \text{and} \quad \tau^2 = \tau_{yz}^2$$

$$\therefore \quad \tau^2 = \tau_{yz}^2 = \frac{T^2(y^2 + z^2)}{J^2}$$

and

$$\tau_{xy} = \tau_{xz} = 0$$

Figure 3-3 Shear Stress Components

Member Strain Energy

The expression for the normal stress and the shear stress distributions from Figures 3-2 and 3-3 are now substituted into the equation for the incremental strain energy for a state of plane strain. A factor K is introduced in the torsion term to account for cross sections that are not circular. The incremental volume dV for which this equation is valid can be expressed as a dA element of the cross section times dx as illustrated in Figure 3-4.

Design for Deflection

Substitue normal and shear stresses into the general incremental plain strain energy equation (dU)

$$dU = \left[\frac{1}{2E}\left\{\left(\frac{P}{A}\right)^2 + \left(\frac{M_z}{I_z}y\right)^2 + \left(\frac{M_y}{I_y}z\right)^2\right\} + \frac{K}{2G}\left\{\frac{T^2}{J^2}(z^2 + y^2)\right\}\right]dAdx$$

(where K is a correction to account for non circular sections)

Figure 3-4 Incremental Member Strain Energy

To obtain the total strain energy of the member this incremental expression must be integrated over the cross sectional area and the length of the member, L. The final equation in Figure 3-5 represents this integration sequence recognizing that when integrating over the cross sectional area the $(P/A)^2$, $(M_z/I_z)^2$, $(M_y/I_y)^2$ and the $(T/J)^2$ are constant over the cross section and can be taken outside of the integration with respect to dA. However, all of these squared loading terms can be functions of x so that integration is left unfinished.

Total strain energy (U) of member is obtained by integrating over the cross section and length of the member.

$$U = \frac{1}{2E}\int_0^L \left\{\left(\frac{P}{A}\right)^2 \int^A dA + \left[(M_z/I_z)^2 \int^A y^2 dA\right] + \left[(M_y/I_y)^2 \int^A z^2 dA\right]\right\} dx$$
$$+ \frac{K}{2G}\int_0^L \left\{\frac{T^2}{J^2}\int^A (z^2 + y^2)dA\right\} dx$$

Figure 3-5 Total Member Strain Energy

Simplified Total Energy

Since the member may be curved the axial x coordinate is replaced by s measured along the curve to account for that difference in geometry. The loading terms may vary along the length of the member. This is

Design for Deflection

recognized by setting $P = P(s)$, $M_z = M_z(s)$, $M_y = M_y(s)$ and $T = T(s)$ in Figure 3-6. It is assumed that the cross section of the member will remain constant along it length. This permits the area integral in the total strain energy expression from the previous slide to be interpreted as follows: the integral of dA over the area is just the cross sectional area A, the integral of $y^2 dA$ becomes I_z the moment of inertia about the z axis, the integral of $z^2 dA$ is equal to the moment of inertia about the y axis, I_y, and the integral of $(z^2 + y^2)$ dA is the polar moment of inertia J of the cross section.

Substituting these area properties into the total strain energy equation and separating the integrals of the loadings with respect to length along the member ds gives the final result. It is again noted that I_y and I_z must be principal moments of inertia of the cross section.

$$P = P(s), \; M_z = M_z(s), \; M_y = M_y(s), \; T = T(s)$$

and $\int^A dA = A, \; \int^A y^2 dA = I_z, \; \int^A z^2 dA = I_y, \; \int^A (z^2 + y^2) dA = J$

so that final simplified form of total strain energy is

$$U = \frac{1}{2EA}\int_0^L \{P(s)\}^2 ds + \frac{1}{2EI_z}\int_0^L \{M_z(s)\}^2 ds + \frac{1}{2EI_y}\int_0^L \{M_y(s)\}^2 ds$$
$$+ \frac{K}{2GJ}\int_0^L \{T(s)^2\} ds$$

Note: I_z and I_y must be principal moments of inertia

Figure 3-6 Simplified Total Energy

Castigliano's Theorem

The development of the strain energy equation in terms of the bending, torsion and axial loadings along

Design for Deflection

the length of the member has been in preparation to take advantage of a powerful method of deflection analysis called Castigliano's theorem. Carlo Alberto Castigliano was an Italian mathematician and physicist who lived in the later half of the 19th century. The theorem that appears in his dissertation simply states that when forces act on an elastic system the deflection at the location of the applied force and in it's direction is given by the partial derivative of the total strain energy with respect to that specific force. This is expressed as δ_i equal to the partial derivative of U with respect to Fi as illustrated in Figure 3-7.

Theorem:

When forces act on elastic systems, the displacement corresponding to any force may be found by obtaining the partial derivative of the total strain energy with respect to that force. That is:

$$\delta_i = \frac{\partial U}{\partial F_i}$$ deflection at location and in direction of F_i

Corrolary:

$$\theta_i = \frac{\partial U}{\partial M_i}$$ slope of member at location and in direction of M_i

Figure 3-7 Castigliano's Theorem

A corollary to the theorem is that the slope of a member at the location of an applied moment is given by the partial derivative of the total strain energy with respect to that specific moment. This is expressed mathematically as theta θi is equal to the partial of U with respect to Mi.

Design for Deflection

Theorem Application

Beginning with the total strain energy in terms of the loadings the application of Castigliano's theorem for finding a specific deflection or slope can be illustrated as shown in Figure 3-8. In the general deflection equation for δ_i the differentiation inside the first integral of the axial force contribution becomes 2 times the axial force, P, times the partial derivative of the axial force with respect to the force F_i at which the deflection is sought. The 2 cancels the one half outside the integral leaving finally 1/AE times the integral of P times the partial of P with respect to F_i integrated with respect to s over the length of the member L. The integrals for the bending and the torsion contributions take on a similar formulation.

This is also true for the general expression θ_i with the exception that the partial derivatives are taken with respect M_i where the slope will be determined.

$$\delta_i = \frac{\partial U}{\partial F_i} = \frac{1}{AE}\int P\frac{\partial P}{\partial F_i}ds + \frac{1}{EI}\int \left(M_y\frac{\partial M_y}{\partial F_i} + M_z\frac{\partial M_z}{\partial F_i}\right)ds + \frac{K}{GJ}\int T\frac{\partial T}{\partial F_i}ds$$

Also

$$\theta_i = \frac{\partial U}{\partial M_i} = \frac{1}{AE}\int P\frac{\partial P}{\partial M_i}ds + \frac{1}{EI}\int \left(M_y\frac{\partial M_y}{\partial M_i} + M_z\frac{\partial M_z}{\partial M_i}\right)ds + \frac{K}{GJ}\int T\frac{\partial T}{\partial M_i}ds$$

Figure 3-8 Theorem Application

Cantilever Beam Example

Castigliano's theorem will now be applied to determine the end deflection and slope of a cantilever beam loaded with a concentrated force F and moment

Design for Deflection

M_o at the free end. The problem will be solved in general and two specific solutions will be examined, the deflection and slope due to F when M_o is zero and the deflection and slope for M_o when F is zero. Measuring x to the left from the free end of the beam the bending moment M as a function of x is given by $M(x) = Fx - M_o$. In Figure 3-9 the partial derivative of $M(x)$ with respect to F is simply x and the partial derivative of $M(x)$ with respect to Mo is just minus one.

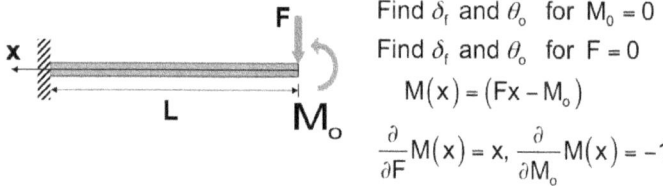

Find δ_f and θ_o for $M_o = 0$
Find δ_f and θ_o for $F = 0$
$$M(x) = (Fx - M_o)$$
$$\frac{\partial}{\partial F} M(x) = x, \quad \frac{\partial}{\partial M_o} M(x) = -1$$

Figure 3-9   Cantilever Deflection Problem

Substituting for $M(x)$ and the derivative of $M(x)$ with respect to F into the general equation for δ gives 1/EI times the integral of the quantity $(Fx - M_o)$ times x dx over the length of the beam L as shown in Figure 3-10.

$$\delta = \frac{\partial U}{\partial F} = \frac{1}{EI}\int M(x)\frac{\partial}{\partial F}M(x)dx = \frac{1}{EI}\int_0^L (Fx - M_o)x\, dx$$

$$\delta = \frac{1}{EI}\left[\frac{Fx^3}{3} - \frac{M_o x^2}{2}\right]_0^L = \frac{1}{EI}\left[\frac{FL^3}{3} - \frac{M_o L^2}{2}\right]$$

Figure 3-10 Deflection due to F and M_o

Carrying out the integration and limit evaluation gives $\delta = (1/EI)((F L^3/3) - (M_o L^2/2))$. The equation for the slope at the end of the beam, θ, is obtained with substitutions of $M(x)$ and the derivative of $M(x)$ with

Design for Deflection

respect to M_o into the last equation in Figure 3-8. The result becomes 1/ EI times the integral over the beam length of the quantity (Fx–Mo) times -dx. Carrying out this integration and limit evaluation in Figure 3-11 results in $\theta = (1/EI)((M_oL - (FL^2/2)))$.

$$\theta = \frac{\partial U}{\partial M_o} = \frac{1}{EI}\int M(x)\frac{\partial}{\partial M_o}M(x)dx = \frac{1}{EI}\int_0^L (Fx - M_o)(-1)\,dx$$

$$\theta = \frac{1}{EI}\left[M_o x - \frac{Fx^2}{2}\right]_0^L = \frac{1}{EI}\left[M_o L - \frac{FL^2}{2}\right]$$

$$\delta_{M_o=0} = \frac{FL^3}{3EI}, \quad \theta_{M_o=0} = -\frac{FL^2}{EI}$$

$$\delta_{F=0} = -\frac{M_o L^2}{2EI}, \quad \theta_{F=0} = \frac{M_o L}{EI}$$

Figure 3-11 Slope due to F and M_o

By setting M_o equal to zero in the general solutions for δ and θ gives the end deflection and end slope for just F as $\delta = FL^3/3EI$ and $\theta = -FL^2/EI$. Both of these results are correct. In a similar fashion the end deflection and slope for just Mo acting on the beam is obtained by setting F equal to zero giving $\delta = -M_o L^2/2EI$ and $\theta = -M_o L/EI$. These results are also correct.

Curved Beam Example –

The problem is to find the horizontal deflection at the end of the curved member in Figure 3-12 at the point where F is applied. The first task is to determine the loading due to F at some cross section a distance s defined by the angle theta from the end of the beam. This requires placing that section of the beam in equilibrium and determining $P(\theta)$, $M(\theta)$ and $V(\theta)$ at that cross section.

Design for Deflection

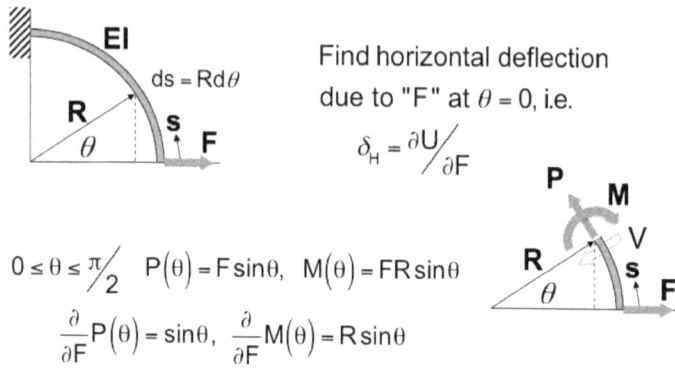

Find horizontal deflection due to "F" at $\theta = 0$, i.e.

$$\delta_H = \partial U / \partial F$$

$0 \le \theta \le \pi/2$ $P(\theta) = F\sin\theta$, $M(\theta) = FR\sin\theta$

$\dfrac{\partial}{\partial F}P(\theta) = \sin\theta$, $\dfrac{\partial}{\partial F}M(\theta) = R\sin\theta$

Figure 3-12 Curved beam Problem

$P(\theta)$ is simply $F(\sin\theta)$ and V is $F(\cos\theta)$. The contribution of V, a shear deflection of the cross section, is neglected since it only becomes important in very short members. The bending moment M due to F is $FR(\sin\theta)$ where R is the radius of the member. The partial derivative of $P(\theta)$ with respect to F is $\sin\theta$ and the partial derivative of $M(\theta)$ with respect to F is $R\sin\theta$.

Substituting the values of $P(\theta)$ and $M(\theta)$ together with their partial derivatives with respect to F into the general equation for δ results in two integral contributions, one due to internal axial loading and a second due to internal bending. The integration will be carried out with respect to ds replaced by $Rd\theta$. It is observed in Figure 3-13 that taking the constant loading terms outside the integral results in the functions to be integrated the integrals are the same for both the axial loading and bending contributions.

The result for δ becomes $FR(1/AE + R^2/EI)$ times the integral of $\sin\theta^2 d\theta$ over the limits of 0 to $\pi/2$.

Design for Deflection

$$\delta_H = \left[\frac{\partial U}{\partial F}\right]_{V=0} = \frac{FR}{AE}\int_0^{\pi/2}(\sin\theta^2)d\theta + \frac{FR^3}{EI}\int_0^{\pi/2}(\sin\theta^2)d\theta$$

$$\delta_H = FR\left[\frac{1}{AE} + \frac{R^2}{EI}\right]\left\{\int_0^{\pi/2}(\sin\theta^2)d\theta\right\}$$

$$\delta_H = FR\left[\frac{1}{AE} + \frac{R^2}{EI}\right]\left[\frac{\theta}{2} + \frac{\sin 2\theta}{4}\right]_0^{\pi/2}$$

$$\delta_H = FR\left[\frac{1}{AE} + \frac{R^2}{EI}\right]\left(\frac{\pi}{4}\right) \quad \text{units}: \text{lb in}\left[\frac{1}{ft^2(lb/in^2)} + \frac{in^2}{(lb/in^2)in^4}\right] = in$$

Figure 3-13 Curved Beam Solution

Performing the integration and limit evaluation results in δ, $= FR(1/AE + R^2/EI)(\pi/4)$. A check of the units of the solution is confirmed.

This result indicates that the horizontal deflection due to F of this quarter curved beam is made up of an axial strain energy contribution as well as bending. This might not have been anticipated and demonstrates both the power and simplicity of the application of Castigliano's theorem to the total strain energy representation. In Chapter 4 a number of additional applications of Castigliano's theorem will be presented to further demonstrate the utility and power of this method of determining deflections including the solution of indeterminate structure problems.

Design for Deflection

Design for Deflection

Chapter 4 – Applications of Castigliano's Theorem

Introduction

Chapter 4 examines the application and results of Castigliano's theorem to a variety of slender beam problems that includes finding deflections at locations where there are no forces as well as indeterminate, multiple member and truss structures.

Cantilever Beam Example

Castigliano's theorem can be used to determine deflections at locations on a structure other than those at external loads. This is accomplished by introducing virtual forces that are subsequently set equal to zero. Consider the problem of determining the vertical deflection at ¾ L from the left end of the cantilever beam illustrated in Figure 4-1.

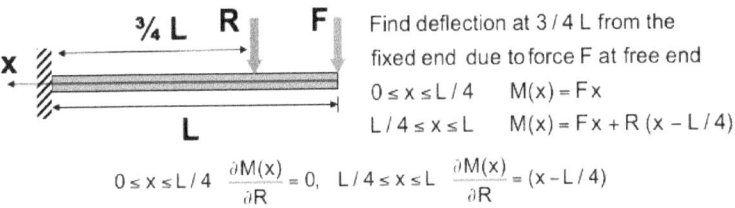

Figure 4-1 Cantilever Beam Example

A virtual force R is introduced at that point. The bending moment M(x) measured from the right

33

end is given by Fx for $0 < x < L/4$ and $Fx + R(x - L/4)$ for $L/4 < x < L$.

The partial derivative of M(x) with respect to R is 0 for $0 < x < L/4$ and $(x - L/4)$ for $L/4 < x$. In Figure 4-2 the deflection at, δ, at R, given by the partial of the total strain energy U with respect to R, consists of two integral covering the two regions of x.

$$\delta_R = \frac{\partial U}{\partial R} = \frac{1}{EI} \int_{L/4}^{L} \left(Fx + R\left(x - \frac{L}{4}\right)\right)\left(x - \frac{L}{4}\right) dx$$

$$\delta_{3/4 L} = \left(\frac{\partial U}{\partial R}\right)_{R=0} = \frac{1}{EI} \int_{L/4}^{L} (Fx)\left(x - \frac{L}{4}\right) dx = \frac{F}{EI} \int_{L/4}^{L} \left(x^2 - \frac{L}{4} x\right) dx$$

$$\delta_{3/4 L} = \frac{F}{EI}\left[\frac{x^3}{3} - \frac{L}{8} x^2\right]_{L/4}^{L} = \frac{FL^3}{EI}\left[\frac{1}{3} - \frac{1}{8} - \frac{1}{192} + \frac{1}{128}\right]$$

$$\delta_{3/4 L} = \frac{324}{1536} \frac{FL^3}{EI} = 0.21 \frac{FL^3}{EI}$$

Figure 4-2 Cantilever Beam Example

For $0 < x < L/4$ the contribution to δ is zero. In the region $L/4 < x < L$ the contribution is 1/EI times the integral from L/4 to L of the moment quantity Fx +R(x - L/4) all times (x - L/4) dx. The value of the virtual force R is then set equal to zero and the integration is carried out and limits evaluated. The final result is 324/1536 FL³/EI. The numerical coefficient of FL³/EI is 0.21, which as expected is less than the deflection coefficient 0.33 at the end of the beam.

Design for Deflection

Uniformly Loaded Cantilever

To determine the end deflection of a uniformly loaded cantilever beam the technique of introducing a virtual fore is again used. The problem solved will be the uniformly loaded cantilever in Figure 4-3 with a concentrated force F at the free end. When the final result for the deflection due to F and the uniform load is obtained F will be set to zero. A change in how this problem will be solved is that the total strain energy will be calculated first before the partial derivative with resect to F is performed. This order of solving the problem will have no effect on the final answer.

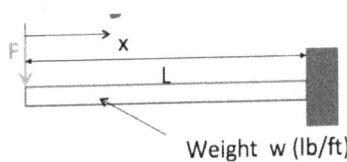

Weight w (lb/ft)
Find end deflection of uniformly weighted cantilever.

Apply force F to location where deflection is desired and proceed with energy method

Bending moment – $\quad M(x) = Fx + \dfrac{wx^2}{2}$

Total strain energy – $\quad U = \dfrac{1}{2EI}\int_0^L \left(Fx + \dfrac{wx^2}{2}\right)^2 dx$

Figure 4-3 Uniformly Loaded Cantilever

The internal bending moment in terms of x measured from the free end of he beam is given by $M(x) = Fx + wx^2/2$. The total strain energy U then becomes 1/2EI times the integral of $M(x)^2 dx$ over the length L of the beam, $0 < x < L$, as illustrated in Figure 4-3.

Design for Deflection

The deflection at the location of F is now determined by taking the partial derivative of U with respect to F. This results in the integral of the quantity $(Fx + wx^2/2)\, x\, dx$ in Figure 4-4.

Determine deflection under end load F

$$\frac{\partial U}{\partial F} = \delta = \frac{1}{EI}\int_0^L \left(Fx + \frac{wx^2}{2}\right)x\,dx = \frac{1}{EI}\int_0^L \left(Fx^2 + \frac{wx^3}{2}\right)dx$$

$$\delta = \frac{1}{EI}\left[\frac{Fx^3}{3} + \frac{wx^4}{8}\right]_0^L = \frac{1}{EI}\left[\frac{FL^3}{3} + \frac{wL^4}{8}\right]$$

Now set F = 0 to get δ for uniformly loaded cantilever

$$\delta_w = \frac{wL^4}{8EI} \quad \Rightarrow \quad \text{correct result}$$

As a check set w = 0 for deflection of end loaded cantilever

$$\delta_F = \frac{FL^3}{3EI} \quad \Rightarrow \quad \text{correct result}$$

Figure 4-4 Uniformly Loaded Cantilever Continued

Carrying out the integration and evaluating it's limits give the result for δ as $1/EI\ (FL^3/2 + wl^4/8)$. Setting F =0 gives $\delta = wlL4/8EI$. This is correct for the end deflection of a uniformly loaded cantilever. If w =0 the deflection $\delta = FL^3/3EI$ which is correct for the end deflection of a cantilever beam with a concentrated end load F.

Curved Beam Example

The problem here is to calculate the horizontal and vertical deflection at the end of the curved beam in Figure 4-5 due to the force F. To find the vertical deflection, δ_V, it is necessary to introduce the virtual fore V. The horizontal deflection, δ_H, will be the

Design for Deflection

partial of the strain energy U with respect to F with V set equal to zero and the vertical deflection will be the partial of U with respect to V with V set equal to zero.

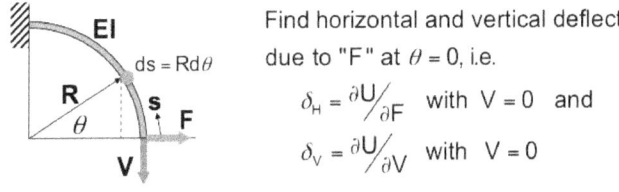

Find horizontal and vertical deflection due to "F" at $\theta = 0$, i.e.

$$\delta_H = \partial U / \partial F \quad \text{with } V = 0 \quad \text{and}$$

$$\delta_V = \partial U / \partial V \quad \text{with } V = 0$$

$0 \le \theta \le \pi/2 \quad P(\theta) = F\sin\theta + V\cos\theta, \quad M(\theta) = FR\sin\theta - V(R - R\cos\theta)$

$$\frac{\partial}{\partial F}P(\theta) = \sin\theta, \quad \frac{\partial}{\partial F}M(\theta) = R\sin\theta$$

$$\delta_H = \left[\frac{\partial U}{\partial F}\right]_{V=0} = \frac{1}{AE}\int_0^{\pi/2}(F\sin\theta)(\sin\theta)Rd\theta + \frac{1}{EI}\int_0^{\pi/2}(FR\sin\theta)(R\sin\theta)Rd\theta$$

Figure 4-5 Curved Beam Example

Equilibrium conditions applied to a section s of the beam defined by the angle θ and located at the dot on the diagram consist of an axial force P(θ) = Fsinθ + Vcosθ and a bending moment M(θ) = FRsinθ - V (R - Rcosθ). The partial of P(θ) with respect to F is (sinθ) and the partial of M(θ) with respect to F is (Rsinθ). The horizontal deflection delta δ_H is equal to the partial of U with respect to F with V set equal to zero. This becomes the axial force contribution of 1/AE times the integral from 0 to π/2 of Fsinθ (sinθ)Rdθ plus 1/EI times the integral from 0 t π/2 of (FRsinθ)(Rsinθ) Rdθ. In both these integrations the incremental length along the beam ds is replaced by Rdθ which accounts for the integral limits being angles.

Design for Deflection

$$\delta_H = \left[\frac{\partial U}{\partial F}\right]_{V=0} = \frac{FR}{AE}\int_0^{\pi/2}(\sin\theta^2)d\theta + \frac{FR^3}{EI}\int_0^{\pi/2}(\sin\theta^2)d\theta$$

$$\delta_H = FR\left[\frac{1}{AE} + \frac{R^2}{EI}\right]\left\{\int_0^{\pi/2}(\sin\theta^2)d\theta\right\}$$

$$\delta_H = FR\left[\frac{1}{AE} + \frac{R^2}{EI}\right]\left[\frac{\theta}{2} + \frac{\sin 2\theta}{4}\right]_0^{\pi/2}$$

$$\delta_H = FR\left[\frac{1}{AE} + \frac{R^2}{EI}\right]\left(\frac{\pi}{4}\right) \quad \text{units}: \text{lb}\,\text{in}\left[\frac{1}{ft^2\left(\frac{lb}{in^2}\right)} + \frac{in^2}{\left(\frac{lb}{in^2}\right)in^4}\right] = in$$

Figure 4-6 Curved Beam Continued

Combining like terms in Figure 4-6 δ_H is given by FR times the quantity {(1AE) + (R²/EI)] times the integral of (sin θ²) dθ from 0 to π/2. Carrying out the integration and evaluating the result at the limits gives a final result of δ_H =FR [(1/AE) +(R2/EI)](π/4). This is the same result obtained at the end of Chapter 3 when the curved beam was only subjected to the force F.

To determine the vertical deflection, δ_V due to the force F the solution begins with the same axial force P(θ) and moment M(θ). The vertical deflection at F is given by the partial of U with respect to V with V set equal to zero. The partial of P(θ) with respect to V in Figure 4-7 is (cosθ) and the partial of M(θ) with respect to V is -(R - Rcosθ). δ_V is then the partial of U with resect to V with V set equal to zero. This consists of a contribution by the axial force distribution and the bending moment distribution.

Design for Deflection

The axial force contribution is (1/AE) times the integral of (F sinθ)(cosθ) Rdθ over the limits of zero to π/2. The bending moment contribution is minus 1/EI) times the integral of (FR sinθ)(R - R cosθ)Rdθ with limits from zero to π/2. The final expression in Figure 4-7 combines those terms that have a common integrand of sinθ cosθ.

$$0 \le \theta \le \pi/2 \quad P(\theta) = F\sin\theta + V\cos\theta, \quad M(\theta) = FR\sin\theta - V(R - R\cos\theta)$$

$$\frac{\partial}{\partial V}P(\theta) = \cos\theta, \quad \frac{\partial}{\partial V}M(\theta) = -(R - R\cos\theta)$$

$$\delta_V = \left[\frac{\partial U}{\partial V}\right]_{V=0} = \frac{1}{AE}\int_0^{\pi/2}(F\sin\theta)(\cos\theta)Rd\theta$$

$$- \frac{1}{EI}\int_0^{\pi/2}(FR\sin\theta)(R - R\cos\theta)Rd\theta$$

$$\delta_V = FR\left[\frac{1}{AE} + \frac{R^2}{EI}\right]\int_0^{\pi/2}(\sin\theta)(\cos\theta)d\theta - \frac{FR^3}{EI}\int_0^{\pi/2}(\sin\theta)d\theta$$

Figure 4-7 Curved Beam Continued

For simplicity the sinθ cosθ term in the first integral is replace by 1/2 sin2θ. Carrying out the indicated integrations and evaluating the result from zero to π/2 yield the final result of (FR/2)[(1/AE) – (R²/EI)} as shown in Figure 4-8. If the axial force contribution is small compared to the bending contribution the deflection will be opposite to the assumed direction of V or vertically upward which is what would be anticipated.

Design for Deflection

$$\delta_V = FR\left[\frac{1}{AE} + \frac{R^2}{EI}\right]\int_0^{\pi/2}\left(\frac{1}{2}\sin 2\theta\right)d\theta - \frac{FR^3}{EI}\int_0^{\pi/2}(\sin\theta)d\theta$$

$$\delta_V = FR\left[\frac{1}{AE} + \frac{R^2}{EI}\right]\left[-\frac{\cos 2\theta}{4}\right]_0^{\pi/2} - \frac{FR^3}{EI}[-\cos\theta]_0^{\pi/2}$$

$$\delta_V = FR\left[\frac{1}{AE} + \frac{R^2}{EI}\right]\left[\frac{1}{2}\right] - \frac{FR^3}{EI} = \frac{FR}{2}\left[\frac{1}{AE} - \frac{R^2}{EI}\right]$$

If axial effect is small then δ_V will be in opposite direction to V and less than δ_H which seems correct

Figure 4-8   Curved Beam Continued

To determine the magnitude of the axial effect on the vertical deflection at F assume that the beam has a solid circular cross section of radius little r. Then the area A will be πr^2 and the moment of inertia I will be $\pi r^4/4$. Substituting these values into the result for δ_V at F results in $\delta_V = (FR/2\pi E)(4R^2/r^4)[(1/4(r/R)^2 - 1)]$ as shown in Figure 4-9.

Assume beam has circular cross section of radius r
which is small compared to R, i.e. $r \ll R$
then with $A = \pi r^2$ and $I = \pi r^4/4$

$$\delta_v = \frac{FR}{2}\left[\frac{1}{AE} - \frac{R^2}{EI}\right] \text{ becomes}$$

$$\delta_v = \frac{FR}{2\pi E}\left[\frac{1}{r^2} - \frac{4R^2}{r^4}\right] = \frac{FR}{2\pi E}\left[\frac{4R^2}{r^4}\right]\left[\frac{1}{4}\left(\frac{r}{R}\right)^2 - 1\right]$$

with $r \ll R$ then $\quad \delta_v \cong -\frac{FR^3}{Er^4}\left(\frac{2}{\pi}\right)$

since axial effect is very small

Figure 4-9   Axial Force Effect

Design for Deflection

If r is small compared to R, say one tenth, then the two terms in the last bracket of the solution are [(1/400) -1]. The contribution from the axial effect is only 0.25 % of the total deflection making the vertical movement upward of magnitude $\delta_V = (FR^3/Er^4)(2/\pi)$.

Now compare δ_V to δ_H. From Figure 4-4 δ_H was determined to be $FR[(1/AE) + (R2/EI)](\pi\pi/4)$. It can be rearranged as shown in Figure 4-10 to $(FR/E)(R^2/r^4)[(1/4)(r/R)^2 + 1]$. The contribution due to the axial component is again small and can be neglected leaving $\delta_H = FR^3/Er^4$. This results in $\delta_V = -0.68\ \delta_H$. The negative vertical displacement is significant even though the force F is applied horizontally. The inclusion of the force V in this analysis and treating it like a virtual force permitted this result to be determined as part of using energy methods to solve this problem.

$$\delta_V = -\frac{FR^3}{Er^4}\left(\frac{2}{\pi}\right) \quad \delta_H = FR\left(\frac{1}{AE} + \frac{R^2}{EI}\right)\left(\frac{\pi}{4}\right)$$

let $A = \pi r^2$ and $I = \frac{\pi r^4}{4}$ then

$$\delta_H = \frac{FR}{E}\left(\frac{1}{\pi r^2} + \frac{4R^2}{\pi r^4}\right)\left(\frac{\pi}{4}\right) = \frac{FR}{E}\left(\frac{R^2}{r^4}\right)\left(\frac{1}{4}\left(\frac{r}{R}\right)^2 + 1\right) \cong \frac{FR^3}{Er^4}$$

so that

$$\delta_V = -\delta_H\left(\frac{2}{\pi}\right) = -0.68\ \delta_H$$

Figure 4-10 Comparison of δ_V to δ_H

Curved Quarter Loop

Consider finding the deflection of the quarter loop in Figure 4-11 in the xy plane loaded by a concentrated force F in the z direction. The internal bending moment M_s acting at a section located by θ indicated by the gray dot is given by FR sinθ. This same section will be subjected to a torsion T_s given by FR (1-cosθ).

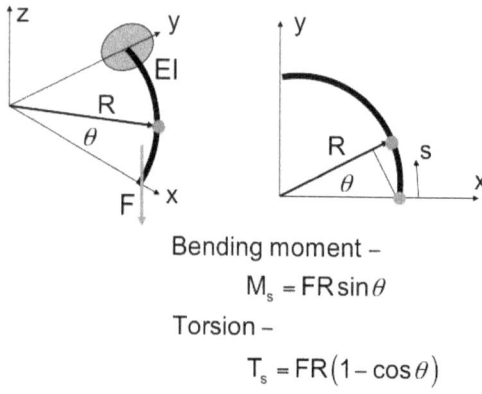

Bending moment –
$$M_s = FR\sin\theta$$
Torsion –
$$T_s = FR(1-\cos\theta)$$

Figure 4-11   Curved Quarter Loop

In this example the total strain energy U, is composed of both a bending and torsional contribution. The strain energy will again be computed before the application of Castigliano's theorem as indicated n Figure 4-12. To perform the integration ds is replaced by Rdθ since the moment and torsion are expressed as functions of theta. The integration limits are zero to π/2.

Carrying out the integration leads to the last equation in Figure 4-12 that needs to be evaluated between the these limits.

Design for Deflection

Strain energy –

$$U = \frac{1}{2EI}\int_0^{R\frac{\pi}{2}} M_s^2 \, ds + \frac{1}{2GJ}\int_0^{R\frac{\pi}{2}} T_s^2 \, ds$$

substitute $\quad ds = R\,d\theta, \quad M_s = FR\sin\theta, \quad T_s = FR(1-\cos\theta)$

$$U = \frac{F^2 R^3}{2EI}\int_0^{\pi/2} (\sin\theta)^2 \, d\theta + \frac{F^2 R^3}{2GJ}\int_0^{\pi/2} (1-\cos\theta)^2 \, d\theta$$

$$U = \frac{F^2 R^3}{2EI}\left[\frac{\theta}{2} - \frac{\theta}{2}\sin\theta\cos\theta\right]_0^{\pi/2} +$$

$$\frac{F^2 R^3}{2GJ}\left[\frac{3\theta}{2} - 2\sin\theta + \frac{\theta}{2}\sin\theta\cos\theta\right]_0^{\pi/2}$$

Figure 4-12   Curved Quarter Loop Continued

Carrying out the limit evaluation leads to the final strain energy U = $(F^2R^3/2EI)(\pi/4)$ + $(F^2R^3/2GJ)$ $(3\pi/4 + 2)$. Differentiating U partial with respect to F and collecting like terms results in the deflection at F as δ = $(FR^3/4)$ $\{(\pi/EI) + [(3\pi + 8)/JG]\}$ in Figure 4-13.

Strain energy –

$$U = \frac{F^2 R^3}{2EI}\left[\frac{\theta}{2} - \frac{\theta}{2}\sin\theta\cos\theta\right]_0^{\pi/2} + \frac{F^2 R^3}{2GJ}\left[\frac{3\theta}{2} - 2\sin\theta + \frac{\theta}{2}\sin\theta\cos\theta\right]_0^{\pi/2}$$

$$U = \frac{F^2 R^3}{2EI}\left(\frac{\pi}{4}\right) + \frac{F^2 R^3}{2GJ}\left(\frac{3\pi}{4} + 2\right)$$

apply Castigliano's therom –

$$\delta = \frac{\partial U}{\partial F} = \frac{FR^3}{4}\left(\frac{\pi}{EI} + \frac{3\pi + 8}{GJ}\right)\left(\frac{lb\,in^3}{(lb/in^2)(in^4)}\right) = (in)$$

Figure 4-13   Vertical Deflection at Force F

This again illustrates that the total strain energy can be calculated before the partial derivative

Design for Deflection

of U with respect to the external load at which the deflection is desired needs to be performed.

To compare the magnitude of the bending and torsion contributions to the deflection in the direction of the force F it will be assumed that the quarter circle section has a solid circular cross section of radius small r. The bending moment of inertia is $I = \pi r^4/4$ and the polar moment of inertia is $J = \pi r^4/2$. It will also be assumed that the shear modulus of elasticity G is approximately one third the value of the modulus of elasticity E.

Substituting these values into the equation for δ and extracting a common term π/EI the two remaining terms in the bracket representing the bending and torsion contributions are 1.0 and 6.29 respectively.

$$\delta = \frac{\partial U}{\partial F} = \frac{FR^3}{4}\left(\frac{\pi}{EI} + \frac{3\pi + 8}{GJ}\right)$$

Assume a circular cross section of radius r

$$I = \frac{\pi r^4}{4} \quad \text{and} \quad J = \frac{\pi r^4}{2}$$

also assume that $G \approx \dfrac{E}{3}$

$$\text{then } \delta = \frac{FR^3}{4}\left(\frac{\pi}{EI} + \frac{3\pi + 8}{GJ}\right) = \frac{FR^3 \pi}{4EI}\left(1 + \frac{\left[3 + \frac{8}{\pi}\right]3}{2}\right)$$

$$\text{or } \quad \delta = \frac{FR^3 \pi}{4EI}(1 + 6.29)$$

Figure 4-14 Bending and Torsion Contributions

This indicates that torsion deformation is significantly larger then the bending effect.

Design for Deflection

Coil Spring Deflection

Now consider the deflection of the quarter-curved section at it center of curvature, point 0, by a concentrated force F as depicted in Figure 4-15.

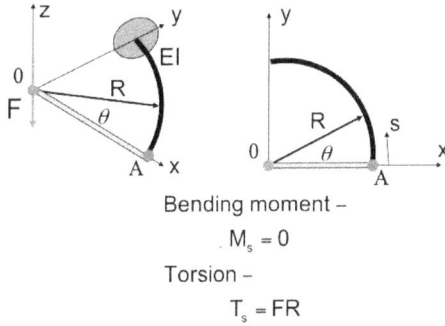

Bending moment –
$M_s = 0$
Torsion –
$T_s = FR$

Figure 4-15 Coil Spring Quarter Loop

Its effect is transmitted to the curved section by mean of a rigid rod from the center to the point A indicated by the dot A shown in Figure 4-13. The bending moment M_s in the section is zero and the torsion T_s is simply the constant FR.

Substituting the bending moment M_s, which is zero, and the torsion T_s =FR into the equation for the total strain energy along with ds = Rdθ results in U equal to $F^2 R^3 /2GJ$ times the integral with respect to θ in Figure 4-16. By integrating from 0 to 2π for θ gives the strain energy for a complete loop as $(FR^3/2JG)(2\pi)$. Differentiating this result with respect to F gives the central deflection for an entire loop as $(FR^3/GJ)(2\pi)$. Now replace J by $\pi d^4/32$ where d is the diameter of the circular cross section of the loop section.

Design for Deflection

Strain energy –

$$U = \frac{1}{2EI} \int_0^{R\frac{\pi}{2}} M_s^2 \, ds + \frac{1}{2GJ} \int_0^{R\frac{\pi}{2}} T_s^2 \, ds$$

substitute $\quad ds = R\,d\theta, \quad M_s = 0, \quad T_s = FR$

$$U = +\frac{F^2 R^3}{2GJ} \int_0^{2\pi} d\theta = \frac{F^2 R^3}{2GJ}(2\pi) \quad \text{for one loop}$$

$$\delta = \frac{\partial U}{\partial F} = \frac{FR^3}{GJ}(2\pi) \quad \text{for circular section } J = \frac{\pi d^4}{32}$$

so $\quad \delta = \dfrac{64 F R^3 n}{G d^4} \quad \left(\dfrac{\text{lb in}^3}{(\text{lb}/\text{in}^2)\text{in}^4} \right) = \text{in} \quad n = \text{number of coils}$

Figure 4-16 Coil Spring Deflections

Multiplying δ by n gives the central deflection of a helical coil spring of n loops with a coil radius of R and a wire diameter of d.

Statically Indeterminate Beam

Castigliano's theorem can be a useful tool in the solution of statically indeterminate beam problems. Consider a cantilever beam loaded with a concentrated force F at it center and simply supported at its free end as shown in Figure 4-17.

This results in a reaction force R at the free end. Since a reaction force and moment are also required at the fixed end the two equations of equilibrium that all external forces and moments acting on the beam must satisfy are insufficient to solve for the two reactions and moment. The problem is statically indeterminate.

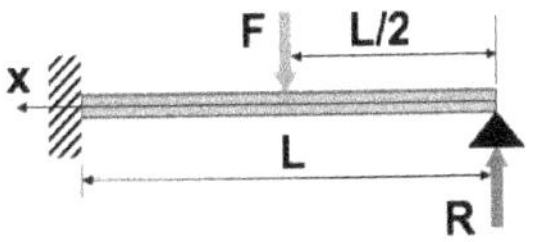

Design for Deflection

Find "R" in terms of "F"

$0 < x < L/2 \quad M_x = -Rx$

$L/2 < x < L \quad M_x = -Rx + F(x - L/2)$

$0 \leq x \leq L/2 \quad \dfrac{\partial}{\partial R}M(x) = -x, \quad L/2 \leq x \leq L \quad \dfrac{\partial}{\partial R}M(x) = -x$

$\delta_R = 0 = \dfrac{\partial U}{\partial R} = \dfrac{1}{EI}\int_0^{L/2}(-Rx)(-x)dx + \dfrac{1}{EI}\int_{L/2}^{L}(-x)\left(-Rx + F\left(x - L/2\right)\right)dx$

Figure 4-17 Statically Indeterminate Cantilever

By recognizing that the deflection at the free end must be zero this provides a condition that permits the magnitude of R to be determined by an application of Castigliano's theorem. This is accomplished by developing an equation for the total strain energy in terms of the forces F and R and then taking the partial of U with respect to R and setting it equal to zero.

The bending moment M(x) for 0 < x < L/2 is simply -Rx. For L/2 < x <L M(x) is –Rx +F(x-L/2). For x <L/2 the partial of M(x) with respect to R is -x. For x > L/2 the partial of M(x) with resect to R is again -x. The deflection δ at R which is zero is given by Castigliano's theorem as 1/EI times the integral from 0 to L/2 of (-Rx) xdx plus 1/ EI times the integral from L/2 to L of - x [(Rx + (F x – L/2)] dx in Figure 4-17.

The last expression from the previous figure is now integrated giving two separate terms in Figure 4-18 that require being evaluated with two different sets of limits. Carrying out the limit evaluations and

47

Design for Deflection

combining like terms leads to the final result that R is equal to 5/16 F. With R determined the reaction and the moment at the wall can be calculated from the two available equations of equilibrium. A second application of Castigliano's theorem could now made if the deflection at the center of the beam was desired by differentiating U partially with respect to F with R equal to 5/16.

$$0 = \left[\frac{Rx^3}{3}\right]_0^{L/2} + \left[\frac{Rx^3}{3} - F\left(\frac{x^3}{3} - \frac{Lx^2}{4}\right)\right]_{L/2}^{L}$$

$$0 = \frac{RL^3}{24} + \frac{RL^3}{3} - F\left(\frac{L^3}{3} - \frac{L^3}{4}\right) - \frac{RL^3}{24} + F\left(\frac{L^3}{24} - \frac{L^3}{16}\right) \Rightarrow$$

$$0 = \frac{RL^3}{3} - \frac{FL^3}{48}(16 - 12 - 2 + 3) = \frac{RL^3}{3} - \frac{5FL^3}{48}$$

or $\quad R = \frac{5}{16}F$

Figure 4-18 Statically Indeterminate Beam (cont.)

Built in Beam at Both Ends

A second example of solving an indeterminate structure problem will be demonstrated here. The problem is to find the deflection at the center of a beam built in at both ends subjected to a concentrated force F as illustrated in Figure 4-17.

Since the problem is symmetric with respect to the center only half of the beam needs be analyzed subject to ½F.

Design for Deflection

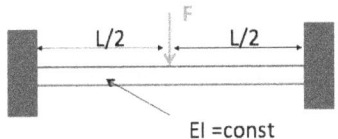

Find deflection at center.
Since problem is symmetric work with only half of beam
and consider free body diagram

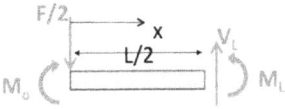

Figure 4-19 Built in Beam at Both ends

A free body diagram of the right half of the beam indicates that a reaction force V_L and moment M_L will exist at the built in end as well as an internal moment M_o at the center shown in Figure 4-20. Since there are only two equations of equilibrium available to solve for these three unknown the problem is statically indeterminate.

The two equilibrium conditions that govern the forces and moments on the beam are applied to determine what relationships exist between these forces and moments.

Summing forces vertically and setting them to zero yields $V_L = F/2$. Summing moments at the right end results in $F - (4M_o/L) + 4ML/L = 0$.

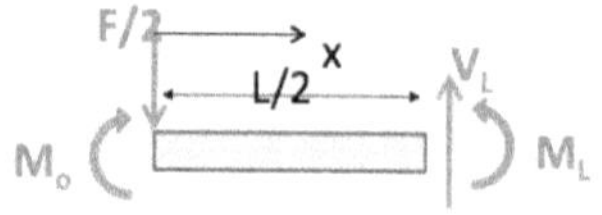

Design for Deflection

$\sum F_x = 0$

$-\dfrac{F}{2} + V_L = 0 \quad \Rightarrow \quad V_L = \dfrac{F}{2}$

$\sum M_L = 0$

$\left(\dfrac{F}{2}\right)\left(\dfrac{L}{2}\right) - M_o + M_L = 0 \quad$ or

$F - \dfrac{4M_o}{L} + \dfrac{4M_L}{L} = 0 \quad$ insufficient to solve for moments

Figure 4-20   Built in Beam at Both Ends (cont.)

This does not permit M_o or M_L to be determined. The physical condition that the slope at the center of the beam must be zero together with Castigliano's theorem will be used to obtain an additional equation involving the moments and the force F. First the total stain energy U is calculated in Figure 4-21. The internal bending moment M(x) with x measured from the center to the right is given by $M(x) = (F/2)x - M_o$. The total strain energy U is given by 2 times the quantity (1/2EI) times the integral from 0 to L/2 of $[(F/2)x - M_o]^2$ dx.

The partial derivative of U with respect to Mo is the slope at the center of the beam and is set to zero. Performing the differentiation, integrating and carrying out the limit evaluation results in the equation that $0 = (FL^2/16) - M_oL/2$. M_o is therefore FL/8 and from Figure 4-20 M_L becomes –FL/8. The negative sign is consistent with what would be expected for the direction of action of M_L at the right end of the beam.

Design for Deflection

Bending moment –

$$M(x) = \frac{F}{2}x - M_o \quad \text{so that}$$

$$U = 2\left\{\frac{1}{2EI}\int_0^{L/2}\left(\frac{F}{2}x - M_o\right)^2 dx\right\}$$

Set slope at center equal to zero, i.e.

$$\frac{\partial U}{\partial M_o} = 0 = \int_0^{L/2}(-1)\left(\frac{F}{2}x - M_o\right)dx$$

$$0 = \left[\frac{Fx^2}{4} - M_o x\right]_0^{L/2} = \frac{FL^2}{16} - \frac{M_o L}{2} \implies M_o = \frac{FL}{8}$$

$$\text{and} \quad M_L = -\frac{FL}{8}$$

Figure 4-21 Built in Beam At Both Ends (cont.)

With the expression for the total strain energy available Castigliano's theorem is now applied to determine the deflection at the center of the beam in Figure 4=22.

δ is the derivative of U with respect to F which is 2/ EI times the integral from 0 to L/2 of x/2 times the quantity [(F/2) -M_o] dx.

Carrying out the indicated mathematical operations results in δ equal to 2/EI times the quantity $(FL^3/96) - (M_o L^2/16)$.

Substituting the value of M_o equal to FL/8 gives a final result for δ at the center of the beam as $FL^3/192\ EI$.

Design for Deflection

$$U = 2\left\{ \frac{1}{2EI} \int_0^{L/2} \left(\frac{F}{2}x - M_o \right)^2 dx \right\}$$

From Castigliano's Theorem

$$\frac{\partial U}{\partial F} = \delta = \frac{2}{EI} \int_0^{L/2} \left(\frac{x}{2} \right)\left(\frac{F}{2}x - M_o \right) dx = \frac{2}{EI} \int_0^{L/2} \left(\frac{F}{4}x^2 - \frac{M_o x}{2} \right) dx$$

$$\delta = \frac{2}{EI}\left[\frac{Fx^3}{12} - \frac{M_o x^2}{4} \right]_0^{L/2} = \frac{2}{EI}\left[\frac{FL^3}{96} - \frac{M_o L^2}{16} \right]$$

Substitute for M_o

$$\delta = \frac{2FL^3}{EI}\left[\frac{1}{96} - \frac{1}{128} \right] = \frac{FL^3}{EI}\left[\frac{1}{48} - \frac{1}{64} \right] = \frac{FL^3}{192EI}$$

Figure 4-22 *Deflection at Center of Built in beam*

Two Element Structure

The crank problem discussed in Chapter 1 will now be analyzed for deflections as a two-element structure using Castigliano's theorem. Recall that for the "s" section the beam is subjected to bending moments M_x and M_y due to F and H. (see Figure 4-23). In the "x" section the bending moments are M_z and M_y. There is also an axial force P as well as a torsion T. For $0 < s < L$ $M_x = Fs$ and $M_y = HL$. For $0 < x < L$ $M_z = Fx$, $M_y = HL$, $P = H$ and $T = FL$.

Over the "s" section the partial of M_x with respect to F is s and the partial of M_y with respect to F is zero. In the "x" section the partial of M_z with respect to F is x, the partial of M_y with respect to F is zero, the partial of P with respect to F is zero and the partial of T with respect to F is just L.

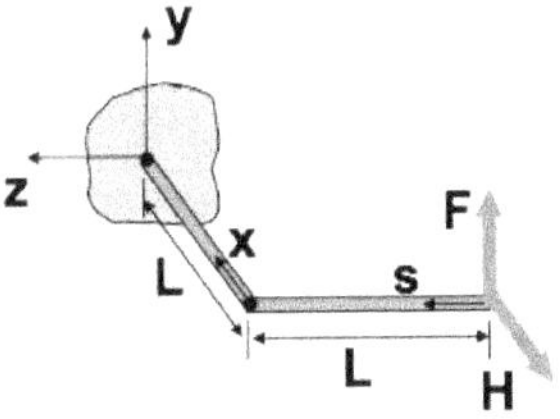

Design for Deflection

Find deflection, δ_F & δ_H

$0 \le s \le L$ $M_x = Fs$, $M_y = Hs$

$0 \le x \le L$ $M_z = Fx$, $M_y = HL$,

$P = H$, $T = FL$

$0 \le s \le L$ $\dfrac{\partial M_x}{\partial F} = s$ $\dfrac{\partial M_y}{\partial F} = 0$, $0 \le x \le L$ $\dfrac{\partial M_z}{\partial F} = x$ $\dfrac{\partial M_y}{\partial F} = 0$ $\dfrac{\partial P}{\partial F} = 0$ $\dfrac{\partial T}{\partial F} = L$

Figure 4-23 Crank Arm Structure

The deflection in the F direction is given by the partial of U the total strain energy, with respect to F (see Figure 4-24). In the "s" section this is given by the integral of (Fs)s ds over the length L. In the "x" section the contribution is the integral of (Fx)x dx over the length L. There is also a torsion contribution of the integral of (FL)L dx. Carrying out these integrations and evaluating the limits results in the final result of $\delta_F = (2\,FL^3/3EI) + (FL^3/GJ)$.

$$\delta_F = \frac{\partial U}{\partial F} = \frac{1}{EI}\left\{\int_0^L (Fs)s\,ds + \int_0^L (Fx)x\,dx\right\} + \frac{1}{GJ}\int_0^L (FL)L\,dx$$

$$\delta_F = \frac{F}{EI}\left\{\int_0^L s^2\,ds + \int_0^L x^2\,dx\right\} + \frac{F}{GJ}\int_0^L FL^2\,dx = \frac{2FL^3}{3EI} + \frac{FL^3}{GJ}$$

Figure 4-24 Crank Arm Structure (cont.)

This result was obtained with a single equation application of Castigliano's theorem as contrasted to the multiple calculations that would be required using a strength of materials approach to analyze each element and superimposing the results.

Design for Deflection

The deflection in the direction of H requires determining the partial of U, the total strain energy, with respect to H. The partial directives of the internal loading with respect to H must be determined first. For the "s" section the partial of M_x with respect t H is zero and the partial of My with respect to H is s. For the x section the partial of Mz with respect to H is zero, the partial of M_y with respect to H is L, the partial of P with respect to H is one and the partial of T with respect to H is zero.

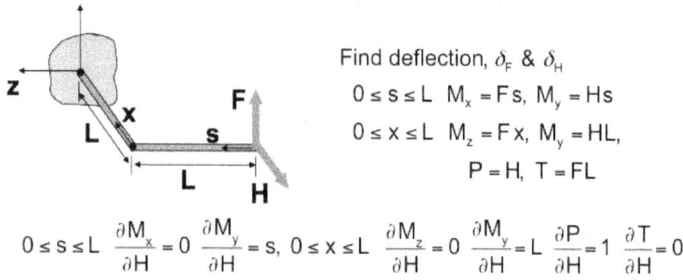

Figure 4-25 Crank Arm Structure (cont.)

The equation for δ_H, equal to the partial of U with respect to H, is made up of two bending contributions, 1/EI times the integral of (Hs)s ds and (HL)L dx over the length of beams and an axial contribution, 1/AE times the integral of H dx over the length L.

The result following integration and limit evaluation is (HL^3) times the quantity [(4/3EI) + (1/AL^2)]. Once again this deflection was obtained with a single equation using Castigliano's theorem.

Design for Deflection

$$\delta_H = \frac{\partial U}{\partial H} = \frac{1}{EI}\left\{\int_0^L (Hs)s\,ds + \int_0^L (HL)L\,dx\right\} + \frac{1}{AE}\int_0^L H\,dx$$

$$\delta_H = \frac{1}{EI}\left\{\int_0^L Hs^2\,ds + \int_0^L HL^2\,dx\right\} + \frac{1}{AE}\int_0^L H\,dx = \frac{HL^3}{EI}\left(\frac{4}{3} + \frac{I}{AL^2}\right)$$

Figure 4-26   Crank Arm Structure (cont.)

Truss Deflection Example

A second multiple element structure amenable to deflection analysis by Castigliano's theorem is a plane truss. To determine deflections of trusses consisting of two force members pinned together subjected to external loadings it is first necessary to determine the axial force in each of the members.

The truss example depicted in Figure 4-27 subjected to the external load F is supported at location A by a pin joint. It will only withstand a horizontal reaction R_1. At point B the pin joint can support both a vertical and horizontal reactions R_2 and V.

From considerations of equilibrium applied to a free body diagram of the entire truss the reaction V is just equal to F, the two reactions R are equal to one another and the reaction R_1 is equal to 2F. At pin A the member force P_{AC} is 2F just balancing R_1.

Design for Deflection

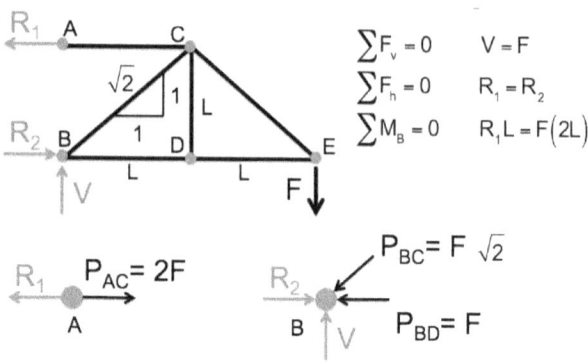

Figure 4-27 Plane Truss Analysis

At pin B in the vertical reaction is balanced by the vertical component of P_{BC} giving P_{BC} a magnitude of $F\sqrt{2}$.

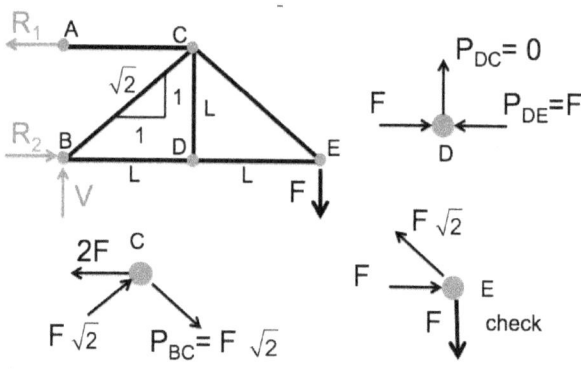

Figure 4-28 Plane Truss Analysis (cont).

At pin D in Figure 4-28 it is observed that P_{DE} is just equal to F so P_{DC} must be zero. At pin C P_{CE} is the only unknown force whose vertical component must be equal to the vertical component of P_{BC} equal

Design for Deflection

to $F\sqrt{2}$. Note that the horizontal components of force at pin C satisfy equilibrium. To check the results equilibrium at pin E is analyzed and found to be satisfied. With the axial forces in all two force members determined the solution for δ due to the external force F can now be determined using Castigliano's theorem.

Truss Deflection at F

The total strain energy of the truss is the sum of the strain energy in each member. This is represented in Figure 4-29 as 1/2AE times the sum of the integrals of P^2 dx over the length of each member of the entire truss.

Since the integral with respect to x is just the length of each member the final generic expression for the total strain energy is given by 1/2AE times the sum from one to m, the total number of members, of P_n^2 squared times L_n. In this particular expression for U it is assumed that AE is the same for each truss member.

For the example problem the total strain energy is made up of three terms The first term is for member AC, $(2F^2)L$. The second term is for members BC and CE, $2(F\sqrt{2})^2(L\sqrt{2})$. The last term is for the members BD and DE, $2(F)^2L$. The total strain energy becomes 1/2AE times the quantity $(2F)^2L + 2(F/\sqrt{2})^2(L\sqrt{2}) + 2F^2L$. Differentiating with respect F gives the deflection δ in the direction of F.

Design for Deflection

$$U = \frac{1}{2AE} \sum_1^m \int_0^{L_n} (P^2)_n \, dx = \frac{1}{2AE} \sum_1^m (P_n)^2 L_n$$

for example problem

$$U = \frac{1}{2AE}\left[(2F)^2 L + 2(F\sqrt{2})^2 L\sqrt{2} + 2(F)^2 L\right]$$

$$U = \frac{F^2 L}{2AE}\left(4 + 4\sqrt{2} + 2\right) = \frac{F^2 L}{2AE}\left(6 + 4\sqrt{2}\right)$$

$$\delta_F = \frac{\partial U}{\partial F} = \frac{FL}{AE}\left(6 + 4\sqrt{2}\right) = 11.8 \frac{FL}{AE}$$

Figure 4-29 Truss Deflection at Force

Design for Deflection

Chapter 5 – Design Related Problems

Introduction

In Chapter 5 two practical design related problems are analyzed and solved using energy techniques and Castigliano's theorem. The first problem deals with a novel method of producing a torsion spring with a cantilever beam. The second asks the question: How high can a very long thin wall cylindrical pipe lying horizontally on the ground be lifted at one point without over stressing the material?

Problem 1 - Cantilever Torsion Spring

A novel design for a torsion spring consists of a cantilever beam whose free end is rigidly attached to a solid disk that can rotate about its center as shown in Figure 5-1. As the disk rotates the beam deflects and a resistance torque is generated that increases with additional rotation of the disk. The geometric design specifications require the disk to be 0.5 inch in diameter and the cantilever to be 3 inches in length. A spring constant of 2 in.lb./deg. is desired. Cross section dimensions for the beam need to be specified.

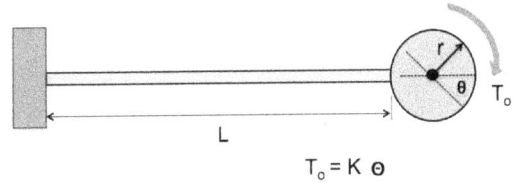

Figure 5-1 Cantilever Torsion Spring

Design for Deflection

Problem Definition

The first task is to determine the spring constant in terms of the geometry of the system and the dimensions of the beam. The resistance torque, represented by T₀, created by the deflection of the beam is a consequence of the beam loading F and M₀ required to produce the dotted deflected beam shape in Figure 5-2.

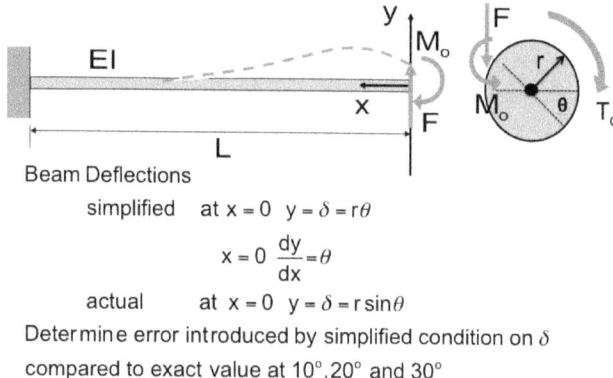

Beam Deflections
simplified   at $x = 0$ $y = \delta = r\theta$

$x = 0$ $\dfrac{dy}{dx} = \theta$

actual   at $x = 0$ $y = \delta = r\sin\theta$

Determine error introduced by simplified condition on δ compared to exact value at 10°, 20° and 30°

Figure 5-2 Beam Deflected Shape and Loading

With the beam rigidly attached to the disk its end will be deflected an amount δ that can be approximated by $r\theta$ provided the rotation θ of the disk is small. At the same time the slope at the end of the beam, dy/dx, will be equal to θ. These two deformations require a vertical force F and a clockwise moment M₀ to be applied to the end of the beam. Their equal and opposite reactions on the disk are balanced by the torque T₀ needed to rotate the disk through the angle θ.

Design for Deflection

As in Figure 5-1 T_o can be written as $T_o = K\theta$ where K is the torsional spring constant of the device. From equilibrium of the disk in Figure 5-2 T_o can also be expressed as $T_o = Fr + M_o$. Hence F and M_o must be determined as a function of θ to permit calculation of the spring constant, K. Castigliano's theorem together with the geometric compatibility of the beam deflection, δ, and slope, dy/dx, with the rotation, θ, will be used to determine F and M_o.

Total Strain Energy

The bending moment M(x) for $0 < x < L$ with x measured from the right end of the beam is $Fx - M_o$. The total strain energy U in the beam is then 1/2EI times the integral of $(FX - Mo)^2$ dx from 0 to L Carrying out the integration and evaluating the result at it's limits gives the final expression for U as $(1/2EI)[(F^2L^3/3) - FMoL^2 + M_o^2L]$, see Figure 5-3.

$$M(x) = Fx - M_o$$

then

$$U = \frac{1}{2EI}\int_0^L (Fx - M_o)^2 dx = \frac{1}{2EI}\int_0^L \left(F^2x^2 - 2FxM_o + M_o^2\right) dx$$

$$U = \frac{1}{2EI}\left[\frac{F^2 x^3}{3} - Fx^2 M_o - M_o^2 x\right]_0^L$$

finally

$$U = \frac{1}{2EI}\left[\frac{F^2 L^3}{3} - FL^2 M_o - M_o^2 L\right]$$

Figure 5-3 Total Beam Strain Energy

Beam Deformations

With the total strain energy determined Castigliano's theorem is applied to obtain equations for δ and dy/dx at the end of the beam in terms of F and M_o. This is carried out in Figure 5-4 by partially differentiating U by F and M_o separately. The two geometric compatibility conditions that must be satisfied between the beam deformations and the rotation of the disk are also listed in Figure 5-4.

$$U = \frac{1}{2EI}\left(\frac{F^2L^3}{3} - FM_oL^2 + M_o^2L\right)$$

apply Castigliano's theorem –

$$\frac{\partial U}{\partial F} = \delta = \frac{1}{EI}\left(\frac{FL^3}{3} - \frac{M_oL^2}{2}\right)$$

$$\frac{\partial U}{\partial M_o} = \frac{dy}{dx_o} = \frac{1}{EI}\left(M_oL - \frac{FL^2}{2}\right)$$

from geometry –

$$\delta = r\theta \quad \text{and} \quad \frac{dy}{dx_o} = \theta \quad \text{use to determine}$$

F in terms of M_o

Figure 5-4 Beam Deformations

Force (F) – Moment (M_o) Relationship

The geometric compatibility conditions are now used to relate the force, F, to the moment, M_o, to produce the deformed beam shape in Figure 5-2. This is carried out in Figure 5-5. For convenience the ratio, r/L, is replaced by λ. With this information the effective spring constant of the devise can now be determined.

Design for Deflection

start with $\quad \delta = r\theta = r\dfrac{dy}{dx}_0$

$$\left(\dfrac{FL^3}{3} - \dfrac{M_oL^2}{2}\right) = r\left(M_oL - \dfrac{FL^2}{2}\right)$$

$$\left(\dfrac{FL^3}{3} - \dfrac{M_oL^3}{2}\right) = \left(M_oL^2\lambda - \dfrac{FL^3}{2}\lambda\right) \quad \text{with } \lambda = \left(\dfrac{r}{L}\right)$$

$$FL\left(\dfrac{1}{3} + \dfrac{\lambda}{2}\right) = M_o\left(\dfrac{1}{2} + \lambda\right)$$

$$F = \dfrac{M_o}{L}\left(\dfrac{\left(\dfrac{1+2\lambda}{2}\right)}{\left(\dfrac{2+3\lambda}{6}\right)}\right) = \dfrac{3M_o}{L}\left(\dfrac{1+2\lambda}{2+3\lambda}\right)$$

Figure 5-5 Force (F) and Moment(Mo) Relationship

Effective Spring Constant

The resistive torque, T_o, is expressed in terms of the spring constant as $T_o = K\theta = K(dy/dx)_o$. From equilibrium of the disk T_o is also $T_o = Fr - M_o$. Combining these two relationships results in the second equation in Figure 5-5 that gives K in terms of F and M_o.

$$T = k\theta = k\left(\dfrac{dy}{dx}_o\right) = Fr + M_o$$

$$\dfrac{k}{EI}\left(M_oL - \dfrac{FL^2}{2}\right) = Fr + M_o = FL\lambda + M_o$$

now substitute for F in terms of M_o

$$\dfrac{k}{EI}\left(M_oL - \dfrac{3M_oL}{2}\left(\dfrac{1+2\lambda}{2+3\lambda}\right)\right) = 3M_o\left(\dfrac{1+2\lambda}{2+3\lambda}\right)\lambda + M_o$$

divide out M_oL and simplify

$$\dfrac{kL}{2EI}\left(\dfrac{2(2+3\lambda) - 3(1+2\lambda)}{(2+3\lambda)}\right) = \left(\dfrac{3(\lambda + 2\lambda^2) + (2+3\lambda)}{(2+3\lambda)}\right)$$

$$k = \dfrac{EI}{L}\left(4 + 12\lambda + 12\lambda^2\right)$$

Figure 5-6 Effective Spring Constant

Design for Deflection

Substituting for F in terms of M_o from Figure 5-5 and eliminating M_o results in a final expression for the effective spring constant K. The effect of the device geometry is included in this equation in terms of the moment of inertia, I, of the beam cross-section and the ratio of r to L

Geometric Approximation Error

Before proceeding to specifying the beam cross section dimensions to provide the required spring constant the approximation for the beam deflection, δ, to the disk rotation, θ, will be investigated. As shown in Figure 5-7 δ is actually $r \sin\theta$ rather than simply $r\theta$. At $\theta = 30°$ the difference in δ is about 5% with the approximation giving the smaller value.

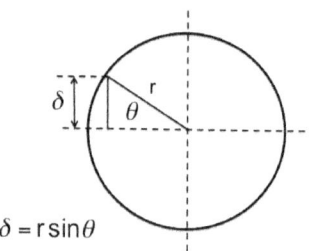

Figure 5-7 Actual Value of Deflection δ

Introducing the correct δ into the geometric compatibility equation in Figure 5-5 results in a transcendental equation between F and M_o as illustrated in Figure 5-8. This does not allow F to be determined directly in terms of M_o.

Design for Deflection

correct δ $\qquad \delta = r\sin\theta = r\sin\left(\dfrac{dy}{dx}\right)_0$

$$\left(\frac{FL^3}{3} - \frac{M_o L^2}{2}\right) = r\sin\left(M_o L - \frac{FL^2}{2}\right)$$

Transandental equation – can not solve for F in terms of M_o directly

Try series expansion for $\sin\theta$, i.e.

$$\sin\theta = \theta - \frac{\theta^3}{3!} + \ldots$$

This will lead to cubic equation – not much help!

Figure 5-8 Attempted Error Correction

Even proposing a series expansion of $\sin\theta$ is of no help as this leads to the requirement of solving a cubic equation in closed form.

Disk Equilibrium Correction

However, a partial correction can be inserted into the disk equilibrium equation that will provide a more accurate solution. This is accomplished by replacing Fr by $Fr\cos\theta$ in the first and second equations in Figure 5-6 as shown in Figure 5-9.

$$T = k\theta = k\left(\frac{dy}{dx_0}\right) = Fr\cos\theta + M_o$$

$$\frac{k}{EI}\left(M_o L - \frac{FL^2}{2}\right) = Fr\cos\theta + M_o = FL\lambda\cos\theta + M_o$$

Figure 5-9 Corrected Disk Equilibrium

The equation for the corrected effective spring constant is determined in Figure 5-10 and checks in the limit with the previous approximate result.

Design for Deflection

now substitute for F in terms of M_o

$$\frac{k}{EI}\left(M_o L - \frac{3M_o L}{2}\left(\frac{1+2\lambda}{2+3\lambda}\right)\right) = 3M_o\left(\frac{1+2\lambda}{2+3\lambda}\right)\lambda\cos\theta + M_o$$

divide out $M_o L$ and simplify

$$\frac{kL}{2EI}\left(\frac{2(2+3\lambda)-3(1+2\lambda)}{(2+3\lambda)}\right) = \left(\frac{3(\lambda+2\lambda^2)\cos\theta+(2+3\lambda)}{(2+3\lambda)}\right)$$

$$k = \frac{EI}{L}\left(4+6\lambda+(6\lambda+12\lambda^2)\cos\theta\right)$$

if $\theta = 0$ then $\cos\theta = 1$ and

$$k = \frac{EI}{L}\left(4+12\lambda+12\lambda^2\right) \quad \text{checks}$$

Figure 5-10 Corrected Effective Spring Constant

It is observed that the corrected value will always be less than the approximate value since $\cos\theta$ decreases as θ increases

Magnitude of Error

The percentage difference between the approximate spring constant, K_a, and the corrected spring constant, K_c, for an angle θ of 30° is calculated in Figure 5-11.

with $r = 0.25"$ and $L = 3"$ $\quad \lambda = .0833$ and $\theta = 30°$

$$K_a = \frac{EI}{L}(4+12\lambda+12\lambda^2)$$

$$K_a = \frac{EI}{L}\left[4+12(.0833)+12(.0833)^2\right] = 5.083\frac{EI}{L}$$

Design for Deflection

and

$$K_b = \frac{EI}{L}\left[4 + 6\lambda + (6\lambda + 12\lambda^2)\cos\theta\right]$$

$$K_b = \frac{EI}{L}\left[4 + 6(.0833) + (6(.0833) + 12(.0833)^2)0.867\right]$$

$$K_b = 5.005\frac{EI}{L}$$

Percent Difference = $\frac{5.083 - 5.005}{5.005} = 1.56\%$

Figure 5-11 Difference between K_a and K_b at $\theta = 30°$

The rate at which this error increases with θ is presented in Figure 5-12. Since the percentage difference is quite small even at $\theta = 40°$ K_a will be used to determine appropriate dimensions for an assumed rectangular cross section of the beam.

Angle	% Difference
0	0
10	0.08
20	0.69
30	1.56
40	3.27

Figure 5-12 % Difference between K_a and K_b

The specified design specifications are r = 0.25 in. and L = 3 in. To proceed further it is first necessary to determine the magnitude of the cross section moment of inertia, I.

Design for Deflection

Calculating Beam Dimensions

In Figure 5-13 the value of I is calculated as 2.26×10^{-6} in^4. For a rectangular cross section and a strip thickness of 0.20" this requires a strip width of 3.38 in. This is obviously too wide as a practical solution.

given $r = 0.25"$, $L = 3"$, $E = 30 \times 10^6$ lb/in^2, $K = 2$ inlb/deg

$$I = \frac{K_a L}{E} \left(\frac{1}{4 + 12\lambda + 12\lambda^2} \right)$$

$$I = \frac{(2)(57.3)(3)}{30 \times 10^6} \left(\frac{1}{4 + 12(0.083) + 12(0.083)^2} \right)$$

$$I = 2.26 \times 10^{-6} \text{ (in}^4\text{)}$$

for a rectangular cross section $I = \frac{(\text{width})(\text{height})^3}{12}$

assume height = 0.020 in. then

$$\text{width} = \frac{(12)(2.26 \times 10^{-6})}{(0.020)^3} = 3.38 \text{ in}$$

Figure 5-13 Calculating Beam Dimensions

The strip width is recalculated for thicker heights, strip thicknesses. The results are tabulated in Figure 5-14

Thickness (in.)	0.020	0.030	0.040	0.050	0.060
Width (in.)	3.380	1.000	0.500	0.220	0.125

Figure 5-14 Calculated Beam Dimensions

From the tabulated dimensions a 0.040" thick x 0.50" wide spring steel strip twill provide the desired torsional spring constant. The 0.050" by

Design for Deflection

0.22" strip would work just as well. Either of these two solutions would appear to be appropriate.

Problem 2 - Lifting a Long Pipe

When long underground pipelines are installed the procedure is to weld finite sections of pipe together and then carefully lower the assembly into a prepared ditch in a manner similar to that depicted in Figure 5-15.

Figure 5-15 Laying Underground Pipeline

With the pipe subjected to significant deformation care must be exercised not to over stress the pipe. To develop an understanding of how the parameters of this system and process affect the design behavior of a long pipe the following problem is proposed for analysis and evaluation.

Design for Deflection

Problem Statement

A very long pipe (effectively infinite in length) of constant cross section and uniform weight per unit length lies on a flat surface. At some point along its length it is raised a vertical distance by the application of a concentrated force. When this takes place a portion of the pipe will rise up off the surface. It is desired to determine a relationship between the magnitude of the force and the distance the pipe is raised as well as the length of the pipe that is no longer in contact with the surface and the maximum bending stress un the pipe.

Apply these results to determine the magnitude of the induced maximum stress, the force required to lift the pipe and the length of pipe lifted from the ground f for a lift of 6 inches in a steel pipe with a diameter of 24" and wall thickness of 3/16".

Problem Definition

Assume that the deformation of this long pipe can be modeled by simple beam theory. Since the pipe is very long and its properties are constant along its length it will deform symmetrically about the point of application of the force that raises it from the surface. Hence, only one half of the structure will be analyzed from the point of force application to where it makes contact with the ground. Assume appropriate external forces and moments and use equilibrium and known conditions of deformation to determine these reactions and the raised length.

Design for Deflection

Free Body Diagram

A free body diagram of one half of the pipe is shown in Fugure5-16. At the right end a force, F, and a moment, M_o, must act to produce the vertical deflection, δ, and maintain a zero slope, $dy/dx = 0$. At $x = L$ where the pipe first lifts from the ground a general unknown reaction, R and moment, M_1, is assumed,

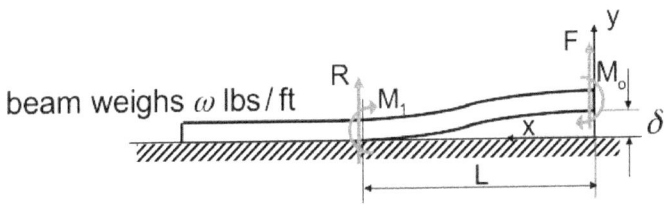

Figure 5-16 FBD of Left Half of Long Pipe

The pipe to the left of the L section lies horizontally on a flat surface. This means that not only is the slope, dy/dx equal to zero but it's second derivative, d^2y/dx^2, is also zero. From this it can be concluded that $M_1 = 0$ since the bending moment cannot change instantly moving into the lifted section. This establishes one of the unknown in Figure 5-16.

Applying Equilibrium

Only two equations of equilibrium are available to determine the remaining forces and moment on the free body diagram. These are represented by equations (1) and (2) in Figure 5-17,

Design for Deflection

vertical force equilibrium and moment equilibrium about the about the left end of the L section.

Vertical equilibrium
$$\sum F_y = 0 \quad F + R - \omega L = 0 \quad (1)$$
Moments about left end of section
$$\sum M_L = 0 \quad FL - M_o - \frac{\omega L^2}{2} = 0 \quad (2)$$
Unknowns:
  F, L, R, M_o
Need two more equations to solve statically indeterminant problem.

Figure 5-17   Equilibrium Equations

Identifying δ as the independent variable in this problem the unknowns become F, R, M_o and L. Two additional equations are required to solve for all unknowns.

Geometric Requirements

Two additional equations involving the unknowns are established from the geometric requirements that at the right end of the pipe the deflection must be equal to δ and the slope dy/dx must be equal to zero as indicated in Figure 5-18.

These requirements will be satisfied by use of the strain energy of bending and Castigliano's Theorem. This is accomplished by first recognizing that the bending moment M(x) can be expressed as $M(x) = Fx - M_o - wx^2/2$ where w is the pipe weight per unit length.

Design for Deflection

At section right end –
  at $x = 0$ $y = \delta$
  at $x = 0$ $\dfrac{dy}{dx} = 0$

Use strain energy and Castigliano's theorem to satisfy these conditions

From the right end –

$$M(x) = Fx - M_o - \frac{\omega x^2}{2}$$

$$U = \frac{1}{2EI}\int_0^L \left[Fx - M_o - \frac{\omega x^2}{2}\right]^2 dx$$

Figure 5-18 Geometric requirements

The integral for the total strain energy in the L section is the last equation in Figure 5-18.

Implementing Geometric Requirements

By differentiating the expression for the total strain energy with respect to F the remaining integral is the deflection δ. Performing this integration and evaluation leads to equation (1) in Figure 5-18 for δ as a function of F, M_o and L.

In a similar fashion the total strain energy integral is differentiated with respect to M_o and set equal to zero. Carrying out the integration and evaluation results in equation (2) in Figure 5-19, another equation relating F, M_o and L.

Design for Deflection

$$U = \frac{1}{2EI}\int_0^L \left[Fx - M_o - \frac{\omega x^2}{2}\right]^2 dx$$

$$\frac{\partial U}{\partial F} = \delta = \frac{1}{EI}\int_0^L \left[Fx^2 - M_o x - \frac{\omega x^3}{2}\right] dx$$

$$\delta = \frac{1}{EI}\left[\frac{FL^3}{3} - \frac{M_o L^2}{2} - \frac{\omega L^4}{8}\right] \quad (1)$$

also $\quad \dfrac{\partial U}{\partial M_o} = 0 = \int_0^L \left[Fx - M_o - \dfrac{\omega x^2}{2}\right] dx$

$$0 = \left[\frac{FL^2}{2} - M_o L - \frac{\omega L^3}{6}\right] \quad (2)$$

Figure 5-19 Implementing Geometric Requirements

F in terms of δ and L

Four independent equations are now available to solve for the four unknowns. By multiplying equation (2) in Figure 5-20 by L/2 and subtracting it from equation (1) results in an expression for F in terms of δ and L

$$\delta = \frac{1}{EI}\left[\frac{FL^3}{3} - \frac{M_o L^2}{2} - \frac{\omega L^4}{8}\right] \quad (1)$$

$$0 = \left[\frac{FL^2}{2} - M_o L - \frac{\omega L^3}{6}\right] \quad (2)$$

Mulitply (2) L/2 and subtract from (1)

$$\frac{EI\delta}{L^3} = F\left(\frac{1}{3} - \frac{1}{4}\right) + \omega L\left(\frac{1}{12} - \frac{1}{8}\right)$$

$$\frac{EI\delta}{L^3} = F\left(\frac{4-3}{12}\right) + \omega L\left(\frac{2-3}{24}\right)$$

so $\quad F = \dfrac{\omega L}{2} + 12\dfrac{EI\delta}{L^3}$

Figure 5-20 F in terms of δ and L

Design for Deflection

Mo in terms of δ and L

The solution for F is now substituted into the second geometric equation, i.e. dy/dx = 0 in Figure 5-18 to determine an equation for M_o in terms of δ and L as shown in Figure 5-21.

substitute $\quad F = \dfrac{\omega L}{2} + 12\dfrac{EI\delta}{L^3}$

into $\quad 0 = \left[\dfrac{FL^2}{2} - M_o L - \dfrac{\omega L^3}{6} \right]$

$0 = \left[\dfrac{\omega L^3}{4} + 6\dfrac{EI\delta}{L} - M_o L - \dfrac{\omega L^3}{6} \right]$

$M_o = \omega L^2 \left(\dfrac{6-4}{24} \right) + 6\dfrac{EI\delta}{L^2}$

$M_o = \dfrac{\omega L^2}{12} + 6\dfrac{EI\delta}{L^2}$

Figure 5-21   Mo in terms of δ and L

Solution for Lifted Length L

The relationships for F and Mo in terms of δ and L are now substituted into the second equilibrium equation in Figure 5-16. This results in an equation in which the only unknown is L as a function of δ.

The other system parameters that effect L are the modulus of the pipe material, E, the cross section moment inertia of pipe, I, and the weight per unit length, w, of the pipe. The final equation for L is given in Figure 5-22.

75

Recall that M(x) is zero at $x = L$, that is

$$M(L) = 0 = FL - M_o - \frac{\omega L^2}{2}$$

with $F = \frac{\omega L}{2} + 12\frac{EI\delta}{L^2}$ and $M_o = \left[\frac{\omega L^2}{12} + 6\frac{EI\delta}{L^2}\right]$

$$M(L) = 0 = \left[\frac{\omega L}{2} + 12\frac{EI\delta}{L^2} - \frac{\omega L^2}{12} - 6\frac{EI\delta}{L^2} - \frac{\omega L^2}{2}\right]$$

$$0 = \left[6\frac{EI\delta}{L^2} - \frac{\omega L^2}{12}\right]$$

so $\delta = \frac{\omega L^4}{72EI}$ or $L = 2.9 \sqrt[4]{\delta EI/\omega}$

Figure 5-22 Length, L, of Lifted Pipe

Determining F, R and M_o

The expression for δ in terms of L is now substituted into the equation for F from Figure 5-19. This permits F to be determined in terms of wL as shown in Figure 5-23.

$$F = \frac{\omega L}{2} + 12\frac{EI\delta}{L^3} = \frac{\omega L}{2} + 12\frac{EI}{L^3}\left(\frac{\omega L^4}{72EI}\right)$$

$$F = \omega L\left[\frac{1}{2} + \frac{1}{6}\right] = \frac{2}{3}\omega L$$

$$R = \omega L - F = \omega L - \frac{2}{3}\omega L$$

$$R = \frac{1}{3}\omega L$$

$$M_o = \frac{\omega L^2}{12} + 6\frac{EI\delta}{L^2} = \frac{\omega L^2}{12} + 6\frac{EI\delta}{L^2}\left(\frac{\omega L^4}{72EI}\right)$$

$$M_o = \omega L^2\left[\frac{1}{12} + \frac{1}{12}\right] = \frac{1}{6}\omega L^2$$

Figure 5-23 Determining F, R and M_o

Design for Deflection

An interesting and surprising result is that F is only 2/3 of the lifted weight of the pipe, rather than the entire weight as might have been expected. A further interesting result is that from the first equilibrium equation the reaction R has a finite value of 1/3 the weight of the lifted pipe. It is not intuitively obvious that a concentrated force should act on the pipe at this location. However, the solution of all the established governing equations gives this result so it must be correct for this mode of pipe deformation. Finally, M_o is also solved for in Figure 5-23 to be $1/6$ wL^2.

Numerical Value of L

To determine the lifted length, L, of a steel 24" pipe with a 3/16 " wall raised 6 " the cross sectional moment of inertia, I, and weight, w, per inch of length must first be calculated. This is illustrated in Figure 5-24. The weight of steel is taken as 480 lb./ft³.

$$I = J/2 = \pi r^3 t = \frac{(3.14)(12)^3(3)}{(16)} = 1017 \text{ in}^4$$

$$A = 2\pi rt = \frac{(6.28)(12)(3)}{(16)} = 14.1 \text{ in}^2$$

$$\omega = w_d A = \left(\frac{480 (\text{lb}/\text{ft}^3)}{1728 (\text{in}^3/\text{ft}^3)}\right)(14.1 \text{ in}^2) = 3.92 \text{ lb/in}$$

$$L = 2.9 \sqrt[4]{\frac{\delta EI}{\omega}} = 2.9 \sqrt[4]{\frac{(6)(30 \times 10^6)(1017)}{(3.92)}} = 13.48 \times 10^2 \text{ in}$$

$$L = \frac{1348}{12} \cong 112 \text{ ft}$$

Figure 5-24 Numerical Value of L

Design for Deflection

Substituting I = 1017 in⁴, w = 3.92 lb./in. and 30 x 10⁶ lb./in.² along with δ = 6 in. into the equation for L results in a value of 112 ft. This may seem a bit long but the pipe is quite stiff due to its large diameter.

Calculation of F and M_o

With L determined as 112 feet for δ of 6 inches the magnitudes of the forces F and R along with the value of M_o can now obtained. Using the equations in Figure 5-22 these calculations are carried out to give F -= 3523 lb., R = 1761 lb. and Mo = 182,000 in. lb.in Figure 5-25.

$$F = \tfrac{2}{3}\omega L = \tfrac{2}{3}(3.92)(1348) = 3523 \text{ lb}$$
$$R = \tfrac{1}{3}\omega L = \tfrac{1}{3}(3.92)(1348) = 1761 \text{ lb}$$
$$M_o = \tfrac{1}{6}\omega L^2 = \tfrac{1}{6}(3.92)(1348)^2 = 1{,}182{,}000 \text{ in lb}$$

Bending stress at pickup point—

$$\sigma = \frac{Mc}{I} = \frac{(1{,}182{,}000)(12)}{(1017)} \cong 13{,}900 \text{ psi}$$

Figure 5-25 Calculation of F, R and Mo

With Mo determined the maximum bending stress is also calculated at the pick up point as 13,900 lb./in².

Shear and Bending Moment Diagram

With the uniqueness of the loading on the L section it is of interest to consider how the internal shear force, V, and the bending moment, M, vary over

Design for Deflection

the region $0 < x < L$. A graph of this variation is presented in Figure 5-26.

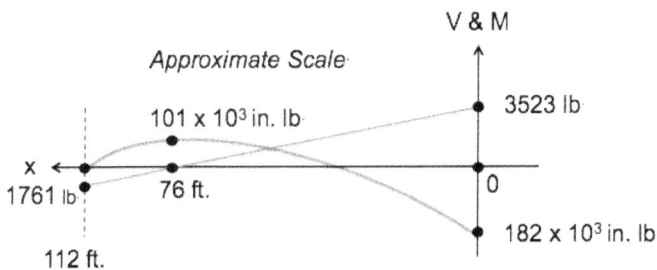

Figure 5-26 Shear and Bending Moment Diagram

The shear decreases linearly from 3523 lb. at $x = 0$ to the negative value of the reaction 1712 ib. at $x = 122$ ft. The moment begins at -182×10^3 in. lb. at $x = 0$ and then increases to a maximum positive value of 101×10^3 in. lb. at $x = 76$ ft. before decreasing rapidly to zero at $x = 112$ ft. where the pipe first lifts from the ground.

The behavior of the bending moment is as expected being positive at the left end of the L section to give a positive upward curvature to the pipe. In a similar fashion the bending moment becomes negative near the point of lifting to provide negative curvature to the pipe so its slope becomes zero at $x = 0$.

The maximum value of the bending moment occurs at $x \mathrel{-}= 0$. This is the location of the maximum bending stress, which is 13,900 lb./in.², the value calculated in Figure 5-25.

Design for Deflection

Total Lifting Force, F_t

The force F calculated in Figure 5-23 is only half of the total lifting force since the section of pipe to the right of $x = 0$ must also be lifted. This makes $F_t = 2F = 4/3$ wL. Combining this value with the equation for δ in terms of L as illustrated in Figure 5-27 results in Ft (lb.) = 4529 $\sqrt[4]{\delta}$ (in.) . For $\delta = 6$ in. Ft =7065

Total Force –

$$F_t = 2F = 2\left(\frac{2}{3}\omega L\right) = \frac{4}{3}\omega L \quad (1)$$

$$\delta = \frac{\omega L^4}{72 EI} \quad (2)$$

Eliminate L between (1) and (2)

$$\delta = \left(\frac{\omega}{72 EI}\right)\left(\frac{3 F_t}{4\omega}\right)^4 = \frac{9}{2048 EI}\left(\frac{F_t^4}{\omega^3}\right)$$

With $E = 30\times10^6$ psi, $I = 1017$ in^4, $\omega = 3.92$ lb/in

$$F_t (lb) = 4529 \sqrt[4]{\delta (in)}$$

For $\delta = 6$ in $F_t = 7065$ lb

Figure 5-27   Lifting Force F_t in terms of δ

Variation of F_t with δ

Without this analysis it might be anticipated that as the lift distance is doubled the force required would also be doubled. However, the relationship between F_t and δ is not linear as indicated in Figure 5-27. What this relationship looks like is presented in the graph in Figure 5-28 where Ft (lb.) is plotted against δ (in.)

Design for Deflection

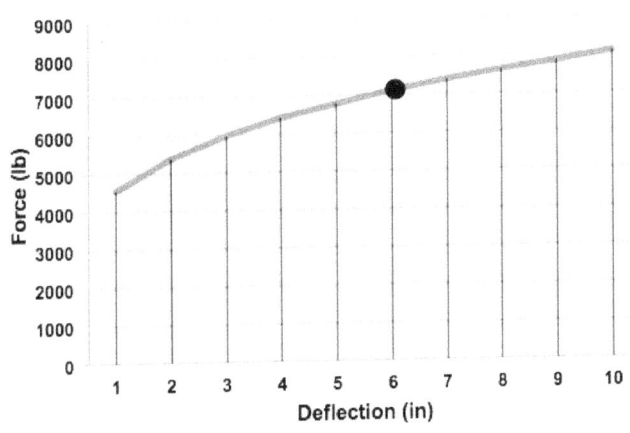

Figure 5-28 Variation of Ft with δ

This very clearly demonstrates the magnitude of the nonlinear behavior of the lifting force to the distance lifted.

Design Parameters in Practical Units

From the photograph in Figure 5-14 it is obvious that practical values of lift are in feet rather than inches. From the information in Figures 5-23, 5-24 and 5-26 the equations for F_t, L, and σ_{max} are with δ measured in inches.

Converting these equations fto δ expressed in feet are presented in Figure 5-29. The units of F_t, L, and σ_{max} remain the same as previously.

Design for Deflection

Length of lifed pipe – L

$$L(\text{ft.}) = 133 \sqrt[4]{\delta(\text{ft.})}$$

Total lift force required – F_t

$$F_t(\text{lb}) = 8420 \sqrt[4]{\delta(\text{ft.})}$$

Maximum bending stress – σ_{max}

$$\sigma_{max}(\text{lb}/\text{in}^2) = 19{,}800 \sqrt{\delta(\text{ft.})}$$

Figure 5-29   L, F_t and σ_{max} in Practical Units

Practical Design Values

The table in Figure 5-30 lists calculated practical operational values for the lifting length, L, (ft.), lifting force, F_t (lb.) and maximum bending stress, σ_{max} (lb./in.²) for vertical lifts, δ from 1 to 6 feet.

Vertical Lift, δ (ft.)	1	2	3	4	5	6
Lifted Length, L (ft.)	133	158	176	188	199	208
Lifting Force (lb.)	8,400	10,000	11,100	11,900	12,600	13,200
Bending Stress (psi)	19,800	27,000	34,300	39,600	44300	48,500

Figure 5-30   Practical Design Values

It is interesting to note that even though the vertical lift is increased by a factor of 6 the lift length and lifting force only increase by an a factor of. 1.5+.

The bending stress increases by a somewhat larger factor of 2.4. This explains why long pipelines can be lifted and moved around as shown in Figure 5-

Design for Deflection

14 without permanently deforming their original lineal geometry.   It also demonstrates how good the value of a complete design operational analysis and evaluation can be.

Design for Deflection

www.ingramcontent.com/pod-product-compliance
Lightning Source LLC
Chambersburg PA
CBHW070103210526
45170CB00012B/734